Seeing what Others Don't

THE REMARKABLE WAYS
WE GAIN INSIGHTS

GARY KLEIN

JOHN MURRAY BUSINESS

This edition published by John Murray Business in 2025
An imprint of John Murray Press

This paperback edition published in 2025

1

A CIP catalogue record for this title is available from the British Library

Paperback ISBN 978 1 39982 660 0
ebook ISBN 978 1 85788 943 7

Printed and bound in Great Britain by Clays Ltd, Elcograf S.p.A.

John Murray Press policy is to use papers that are natural, renewable and recyclable products and made from wood grown in sustainable forests. The logging and manufacturing processes are expected to conform to the environmental regulations of the country of origin.

John Murray Press
Carmelite House
50 Victoria Embankment
London EC4Y 0DZ

John Murray Business
Hachette Book Group
123 South Broad Street
Ste 2750
Philadelphia, PA 19109, USA

www.johnmurraybusiness.com

John Murray Press, part of Hodder & Stoughton Limited
An Hachette UK company

The authorised representative in the EEA is Hachette Ireland,
8 Castlecourt Centre, Dublin 15, D15 XTP3, Ireland (email: info@hbgi.ie)

CONTENTS

PART I

ENTERING THROUGH THE GATES OF INSIGHT

How Do Insights Get Triggered?

CHAPTER ONE

Hunting for Insights

THIS WASN'T SUPPOSED TO BE A MYSTERY STORY. It started out innocently as a collection of clippings from newspapers and magazines. I would come across an article describing how someone made an unusual discovery, and I'd add it to a stack on my desk. The stack included notes describing stories I'd heard during interviews or in conversations. Like other enthusiasms, the stack sometimes got covered up in the competition for space. But unlike the rest, this stack survived. Whenever it got completely buried, it recovered each time I found another article and searched for a place to put it. This pile of clippings endured the occasional bursts of house cleaning that sent many of its neighbors into the purgatory of my file cabinets, if not the trash basket. I'm not sure why it survived. I didn't have any grand plans for it. I just liked adding new material to it. And I liked sifting through it every few months, savoring the stories.

Here's an example of the type of incident that made its way into my stack. Two cops were stuck in traffic, but they didn't feel impatient. They were on a routine patrol, and not much was going on that morning. The older cop was driving. He's the one who told me the story, proud of his partner. As they waited for the light to change, the younger cop glanced at the fancy new BMW in front of them. The driver took a long drag on his cigarette, took it out of his mouth, and flicked the ashes onto the upholstery.

"Did you see that? He just ashed his car," the younger cop exclaimed. He couldn't believe it. "That's a new car and he just ashed his cigarette in that car." That was his insight. Who would ash his cigarette in a brand new car? Not the owner of the car. Not a friend who borrowed the car. Possibly a guy who had just stolen the car. As the older cop described it, "We lit him up. Wham! We're in pursuit, stolen car. Beautiful observation. Genius. I wanted to hug him it was so smart."

I like this kind of story that shows people being clever, noticing things that aren't obvious to others. They're a refreshing antidote to all the depressing tales in the popular press about how irrational and biased we can be. It feels good to document times when people like the young police officer make astute observations.

What changed the fate of this stack of discoveries was that I couldn't answer an important question. I am a cognitive psychologist and have spent my career observing the way people make decisions. Different types of groups invite me to give talks about my work. In 2005, I learned about a movement called "positive psychology," which was started by Martin Seligman, a psychotherapist who concluded that his profession was out of balance. Therapists tried to make disturbed and tormented people less miserable. However, eliminating their misery just left them at zero. What about the positive side of their experience? Seligman was looking for ways to add meaning and pleasure to the lives of his clients.

I felt that the concept of positive psychology applied to decision making as well. Decision researchers were trying to reduce errors, which is important, but we also needed to help people gain expertise and make insightful decisions. Starting in 2005, I added a slide to my presentations showing two arrows to illustrate what I meant. Here is an updated version of that slide:

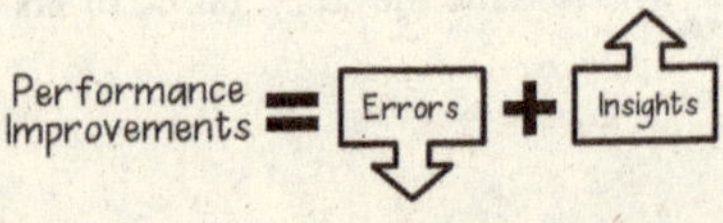

To improve performance, we need to do two things. The down arrow is what we have to reduce, errors. The up arrow is what we have to increase, insights. Performance improvement depends on doing both of these things.

We tend to look for ways to eliminate errors. That's the down arrow. But if we eliminate all errors we haven't created any insights. Eliminating errors won't help us catch a car thief who chooses the wrong moment to flick his ashes.

Ideally, reducing mistakes would at least help us gain insights but I don't believe that's how it works. I suspect the relation between the arrows runs the other way. When we put too much energy into eliminating mistakes, we're less likely to gain insights. Having insights is a different matter from preventing mistakes.

When I showed this slide in my seminars, I got a lot of head nods. The participants agreed that their organizations were all about the down arrow. They felt frustrated by organizations that stifled their attempts to do a good job. Their organizations hammered home the message of reducing mistakes, perhaps because it is easier for managers to cut down on mistakes than to encourage insights. Mistakes are visible, costly, and embarrassing.

However, I also started getting a question: "How can we boost the up arrow?" The audiences wanted to know how they could increase insights. And that was the question I couldn't answer. How to boost insights? I had to admit that I didn't know anything about insights. This admission usually drew a sympathetic laugh. It also drew requests to come back if I ever learned anything useful about insights.

After one such seminar in Singapore, I had a long flight back to the United States to reflect on the up arrow. I wished I could help all the people who wanted to restore a balance between the two arrows in the equation. And then I remembered my stack of clippings that was waiting for me back home.

So in September 2009, I started my own investigation of insight. I began collecting more examples. I was just poking around, nothing serious.

I wanted to explore how people come up with unexpected insights in their daily work. Most studies on insight take place in laboratory settings using college undergraduates trying to solve artificial puzzles. I wondered if I could learn anything useful by studying the way people form insights in natural settings.

I didn't anticipate that this project was going to dominate my attention for the next few years. I didn't foresee that I was going to get drawn into a mystery.

Actually, I got drawn into two mysteries. The first was, What sparks an insight? What happens that lets us make sense of a jumble of unconnected and sometimes contradictory facts, events, and impressions?

Once I got going on that one, a second mystery arose (covered in Part II): What prevents us from grasping an insight? Even when it sits dangling in front of our eyes, ripe for the plucking? Even when others brighten at what they have unexpectedly uncovered?

As I tried to sort that mystery out, I began wrestling with a third issue, more of a challenge than a mystery: Are there any practical ways to increase the flow of insights? That's what my audiences wanted to know, and we'll come to it in Part III. But I'm jumping ahead. At the start, I just wanted to get a better sense of what happens when people have insights. Here are a few of the stories from my collection.

LIGHTING UP LIFE

Martin Chalfie is a soft-spoken man with a relaxed way of describing complicated topics. He is a professor in the Biological Sciences Department at Columbia University, conducting research on the nervous system of worms. One day, almost twenty-five years ago, he walked into a casual lunchtime seminar in his department at Columbia to hear a lecture outside his field of research. An hour later he walked out with what turned out to be a million-dollar idea for a natural flashlight that would let him look inside living organisms to watch their biological processes in action. Chalfie's insight was akin to the invention of the microscope, enabling researchers to see what had previously been invisible. In 2008, he received a Nobel Prize in Chemistry for his work.

You can tell that he's still a bit surprised at the way things worked out for him. He majored in biochemistry at Harvard, but after a disastrous summer laboratory experience at the end of his junior year, he soured on the notion of becoming a researcher. He finished the requirements for his major in his senior year but mostly took courses in law, theater, and Russian literature. He didn't know what he was going to do after college. After graduating in 1969, he worked selling dresses in his parents' dress manufacturing business and also taught at a day school in Connecticut. But when one of his old laboratory projects turned into a publication, he gained the confidence to apply to graduate school at Harvard, and he completed his PhD in 1977.

When the Nobel Prize Committee called him in October 2008, he was sleeping and never heard the phone ring. Later, when he finally woke up, he knew that the prize was to be awarded that day, and in the absence of any notification, he assumed someone else had won. He said to himself, "Okay, who's the schnook that got the Prize this time?" He opened his laptop, went to the Nobel Prize site, and discovered that he was the schnook.

Something happened to him during the hour he spent at the seminar that started his path to the Nobel Prize. Chalfie was studying the nervous system of worms. The type of worms he investigated just happened to have translucent skin, an incidental feature that had played no part in his project up to that point. To study the neurons of worms, Chalfie's assistants had to kill the worms in order to examine their tissues. Chalfie hadn't given the methodology for running these experiments much thought because it was the standard way for researchers like him to do their work.

The speaker at the April 25, 1989, lunchtime talk, one of the regular Tuesday seminars arranged by his department, covered a range of issues that didn't particularly interest Chalfie. Then, in the middle of the talk, the speaker described how jellyfish can produce visible light and are capable of bioluminescence. In 1962, a Japanese scientist discovered the protein that fluoresces to produce a green light in the jellyfish. When ultraviolet light is shined on the protein, it reacts by emitting green light.

That was Chalfie's eureka moment. Suddenly, he understood that if he inserted the green fluorescent protein (GFP) into his transparent worms, he could shine ultraviolet light on it and see where the protein was spreading. He could track the cells into which he placed the GFP. He thought, "I work on this transparent animal, this is going to be terrific! I'll be able to see the cells within the living animal."

Chalfie doesn't remember much about the rest of the lecture because he was so busy making notes about how he could use this GFP as a biological flashlight.

Today, these biological flashlights are a workhorse of molecular biology and a multimillion-dollar industry. Other researchers cloned the GFP so that technicians don't have to chop up lots of jellyfish to extract it. The GFP now comes in additional colors, such as blue, cyan, and red. The GFP is easily inserted into a variety of organisms, not just jellyfish and worms, and it has been put to all kinds of uses. When scientists add the GFP to a virus that is injected into mice, they can watch the virus spread and interact with the immune system. Cancer researchers have inserted the GFP into viruses that grow inside prostate cancer cells, making the physiology of these cells visible. The GFP can be added to a molecule that binds to nerve cells so that surgeons can illuminate nerve fibers that they might otherwise have cut by mistake.

The protein has other important uses. One is detecting pollution. When inserted into a bacterium, the GFP glows brighter as pollution levels increase. Another use is for agriculture. Farmers no longer have to spray an entire field. Instead, they can track which plants the insects are attacking and spray only that part of the field. Some technologists have wondered if it is practical to grow bioluminescent trees that could replace streetlights, thereby reducing shadows and cutting energy costs. There was a puppy named Ruppy (short for Ruby Puppy). She was a cloned beagle, the world's first transgenic dog, and she glowed ruby-red when the protein was activated.

Chalfie's insight about luminescence shows some classical features of the way ideas fit together to form insights. His discovery came without warning. It was emotional, a sudden jolt of excitement. It emerged from a combination of ideas—the transparent worms and the protein that emitted green light. His insight transformed his direction. Before Chalfie walked into the seminar, his investigation of the worm neurons was central to his work and the methods were the background. When he walked out of the seminar, his ideas for a new method took center stage.

No one else in the lunchtime audience had this insight because only Chalfie was studying a transparent animal. And the insight was an act of creation that resulted in something new—Chalfie could use the green fluorescent protein to watch the workings of neurons in a living organism.

If we had an insight Geiger counter, these cues would set it off: a sudden discovery, a jolt of excitement, a combination of ideas that fit tightly together, a feeling of confidence in the new direction. And no one else has the insight, despite receiving the same information. These cues tell us that an insight has just appeared. They are like the green light that Chalfie used to trace living processes.

SPOTTING A MASTER SWINDLER

Bernie Madoff ran the largest Ponzi scheme in history before being arrested in 2008. But a side story described an obscure financial analyst, Harry Markopolos, who way back in 1999 became convinced that Madoff was dishonest. For the next decade, Markopolos set off on Madoff's trail like Inspector Javert in *Les Misérables* trying to bring Jean Valjean to justice, except in this case, Madoff was guilty of more than stealing a 40-sou coin from a young chimney sweep. Like Javert, Markopolos was dogged in his pursuit of Madoff, offended by the thought that a lawbreaker should walk free.

Markopolos notified the Securities and Exchange Commission (SEC) about Madoff in 2000, and his warnings continued until 2008 when Madoff turned himself in. Markopolos got the brush-off each time. The SEC kept treating him as a crank because Madoff was highly respected, a former NASDAQ chairman and well-connected philanthropist. Madoff had sat on the Board of Directors of the Securities Industry Association. No one had heard of Markopolos, who was rumpled where Madoff was smooth, excitable where Madoff was calm. Markopolos himself admits that he is a bit eccentric—for example, naming his twin sons Harry Louie and Louie Harry. More seriously, you have to be a bit nuts to embark on a prolonged investigation the way Markopolos did.

Markopolos's credibility wasn't helped by the fact that he was a competitor of Madoff's in the financial services industry. In addition, he hinted to the SEC about getting a reward. The Securities and Exchange Commission had reasons to be suspicious of Markopolos's accusations.

Markopolos has his own explanations for the brush-off. The SEC bureaucracy isn't well designed to catch frauds of this magnitude, and the SEC staff members don't have the skills to pursue a complicated fraud. Markopolos also believes that government agencies such as the SEC are more interested in protecting Wall Street than in investigating it.

In Part II, we'll explore the reasons that organizations such as the SEC stifle insights. My initial interest in Markopolos was that he recognized from the start that Madoff was dishonest. How did he do it?

In 1999, Markopolos was working at Rampart Investment Management Company in Boston. Frank Casey, one of Markopolos's colleagues, challenged him to match the outstanding results of Bernie Madoff's investment firm. Markopolos was skeptical that anyone could achieve such consistent rates of return, but he agreed to study Madoff's success. And there is another detail to this example: Markopolos was also a certified fraud examiner.

Initially, Markopolos was just curious about how Madoff was operating. "We weren't looking for a crime; we simply wanted to see how he made his numbers dance." He got hold of the publicly available data on Madoff's hedge fund and within minutes knew something was wrong. The numbers just didn't add up. Madoff was said to be using a conservative strategy that wasn't designed to yield consistent profits. Yet Madoff was claiming that his investments were profitable month after month. In fact, Madoff reported losing money in only three months over a period of seven years.

In his book *No One Would Listen,* Markopolos describes his reaction when he first saw a sheet of Madoff's results:

> I glanced at the numbers . . . and I knew immediately that the numbers made no sense. I just knew it. I began shaking my head. I knew what a split-strike strategy was capable of producing,* but this particular one was so poorly designed and contained so many glaring errors that I didn't see how it could be functional, much less profitable. At the bottom of the page, a chart of Madoff's return stream rose steadily at a 45-degree angle, which simply doesn't exist in finance. Within minutes I told Frank, "There's no way this is real. This is bogus."

The odds were astronomical against Madoff reliably sustaining the rate of return he had claimed for so many years.

Markopolos didn't know how Madoff was cheating, although he suspected that Madoff was illegally misdating the times that he placed orders. The other explanation, that Madoff could be running a Ponzi scheme, seemed too far-fetched.

The Markopolos insight that Madoff had to be cheating was sudden, just like those of Chalfie and the young cop. Markopolos used his experience as a fraud investigator to spot telltale implications that others didn't pick up on. Implications that were striking to Markopolos, Chalfie, and the cop were invisible to people without their background and training.

Markopolos, Chalfie, and the young cop all transformed their thinking. After arriving at the insights, they held a different set of beliefs than the ones they had started with. In Markopolos's case, the insights contradicted

* A split-strike conversion strategy involves buying a basket of stocks as well as option contracts on them. For each stock the investor buys a "put" option to protect against the price falling too low and at the same time sells a "call" option to let someone else buy it if it rises above a given "strike" price. If the price rises, the investor makes a profit but only up to the strike price—the gain is capped. If the price falls far enough that it hits the put level, the investor has limited the loss.

his original beliefs. Before he looked at the numbers, Markopolos couldn't imagine that a man with such renown and celebrity as Madoff could engage in a crude swindle. After he looked at the numbers, Markopolos wondered how Madoff was pulling off his fraud.

Yet the stories of these three men differ in important ways. Chalfie noticed how different ideas fit together. Markopolos and the young cop each noticed that some data points did *not* fit together. Chalfie's insight was about how he could build on a combination of ideas. Markopolos and the cop had insights that certain beliefs were unlikely, if not impossible.

Right away, as I studied the incidents in my stack, I could see lots of differences among them and I doubted that I'd find a common script for how all of these insights worked. Here is a fourth incident.

STUMBLING ONTO A PLAGUE

Michael Gottlieb, MD, published the first announcement of the acquired immune deficiency syndrome (AIDS) epidemic. After receiving his medical degree, he did a fellowship at Stanford University on the immune system. In 1980, Gottlieb started out as an assistant professor at UCLA studying the effect of radiation on the immune system of mice. He didn't find this type of research very captivating and was on the lookout for patients with interesting conditions. In January 1981, a young resident told Gottlieb about an unusual case—a thirty-one-year-old man with a yeast infection in his throat. The severe infection made it difficult for the man to breathe. Gottlieb knew that this condition typically affected people who had defective immune systems, and accepted the patient.

Gottlieb tested a sample of the patient's blood. The results didn't make sense. Our immune systems contain different kinds of white blood cells. *Helper* cells activate the immune reaction by triggering disease-fighting cells and guiding the body to produce antibodies that destroy microbial invaders. *Suppressor* cells keep the immune system in check. We have more helper cells than suppressor cells, particularly when we get sick. But this patient was just the reverse. He had more suppressor cells than helper cells. In fact, the patient had hardly any helper cells. Whatever was wrong with him, it was destroying only one type of white blood cell, his helper cells. Gottlieb couldn't find any way to explain these results.

When the patient developed a fever and pneumonia several days later, Gottlieb made arrangements to assay his lung tissue. The patient had *Pneumocystis* pneumonia, a disease caused by a fungus that attacks the fibrous lining of the lungs and interferes with the transport of oxygen into the blood. This yeastlike fungus is sometimes found in the lungs of healthy people. However, it rarely gets out of control unless something goes wrong with a person's immune system. *Pneumocystis* pneumonia affects cancer patients, people receiving organ or bone marrow transplants that require drugs to suppress their immune systems, premature infants, and the elderly. Healthy young adults don't get it. So Gottlieb had another piece of evidence that this patient had something wrong with his immune system.

Gottlieb remembered other things about his patient. The attractive young man was a model who'd even had cheekbone implants. Gottlieb's patient had moved to Los Angeles to live a gay lifestyle. Gottlieb overheard a telephone call in which the patient confided in a friend, "These doctors tell me that I am one sick queen." Such candor, while common today, was unsettling thirty years ago. Like most people, Gottlieb wasn't used to it.

A few months later, Gottlieb examined two others in the Los Angeles area with some of the same symptoms. Both also came down with *Pneumocystis* pneumonia. Gottlieb saw the similarity to his earlier patient and noticed a coincidence: these two men were also gay.

By April, Gottlieb had his fourth and fifth *Pneumocystis* pneumonia patients, with all the typical symptoms: swollen lymph nodes, fever, weight loss, and a nasty case of yeastlike fungal infection. Like the others, these men were gay.

To get the word out quickly, Gottlieb and his colleagues published their findings in the *Morbidity and Mortality Weekly Report,* issued by the Centers for Disease Control. That paper was the first public announcement of the beginning of an epidemic that came to be called "AIDS." Gottlieb's paper,

"*Pneumocystis* Pneumonia—Los Angeles," appeared on June 5, 1981.

Gottlieb's insight centered on a frightening pattern. He didn't know what caused this coincidence—his insight didn't extend that far. He just knew that the cluster of cases seemed ominous. In December, Gottlieb had no inkling of the onrushing AIDS epidemic. By May, he was sounding the alarm. His belief system had been profoundly transformed. So had his medical practice; he began specializing in working with AIDS patients. Years later Gottlieb was Rock Hudson's physician when the actor was first diagnosed with AIDS. Hudson, a six-foot, five-inch romantic actor, had often been voted the favorite leading man by film magazines. He was the first major celebrity to die from AIDS, giving the disease a face that the public could recognize.

Gottlieb built his career around his discovery of AIDS. He published more than fifty papers on AIDS in the mid-1980s and was an investigator on the early clinical trials of the HIV-suppression drug AZT. He was one of the founding chairs of the American Foundation of AIDS Research, a charity established through a $250,000 gift from Rock Hudson's estate. Later, Gottlieb's celebrity status was tarnished when the Medical Board of California reprimanded him and two other physicians for overprescribing painkillers for Elizabeth Taylor, another founder of the charity.

Unlike the first three examples, Gottlieb's insight transformation wasn't sudden. It grew from case to case. What began as a curiosity in January turned into a suspicion in February when Gottlieb saw the second and third AIDS patients, then transformed into a pattern with the fourth and fifth patients. Gottlieb's insight was to see the pattern, as opposed to Chalfie, who spotted an opportunity to combine seemingly unrelated ideas, and Markopolos and the young cop, who both homed in on an inconsistency.

GETTING MY CAR FIXED

Insights aren't reserved for people who win Nobel Prizes, sound the alarm about master criminals, or unravel the mystery of new epidemics. People have insights all the time. Sometimes we notice them, as in the story about the young cop who spotted a car thief. Usually, they're so trivial we don't pay much attention to them unless we're collecting them as a hobby. For example, I was once scheduled to drop off my car for service on a Monday, but that afternoon my mechanic, Don Friessen, telephoned that he was backed up and wouldn't be able to work on it until Wednesday. Unfortunately, it was the Wednesday before Thanksgiving and my wife Helen and I were driving her car out of town that morning. I didn't want to leave my car at Don's repair shop all weekend because then my house might look deserted with no cars in the driveway, possibly attracting the wrong kind of attention.

I told Don that when he finished working on my car, I'd like him to bring it back to my home, a five-minute drive from his shop. I have been going to him for several decades, so he readily agreed even though it would mean more work for him—arranging for someone to drive down separately and pick him up once he dropped off my car.

Then a few hours later, while Helen and I were at a restaurant, I realized there was a better solution. I could just drop off a spare key with Don. When he was ready to work on my car, he'd drive his truck to my house, swap the truck for my car, and then reverse the process when he finished. That way we would have a truck or car in the driveway at all times and Don wouldn't have to make any special arrangements.

Unlike the Chalfie, Markopolos, and cop examples, my little insight came after a period of incubation. During dinner it just popped into my head without any new information. My routine with Don until that day was to bring my car to him. Helen would pick me up after I dropped off the car. Then when it was ready, Don would bring it to me and I'd run him back to his shop. I had never thought about the transaction any other way. I had never considered the advantage of turning a car drop-off into a key drop-off.

There was no great creativity involved in coming up with this solution. It's simply an illustration that everyday insights are much more common than we might think.

We all have a natural tendency to gain insights. We're on the lookout for patterns, as Gottlieb was, and see connections and associations that might be important, just as Chalfie did. We notice inconsistencies, like the young cop and like Markopolos, getting suspicious when we spot irregularities that might be important. We get excited when we find better ways to do things, as I did with my car keys, or when we find new opportunities, like Chalfie. Many people spend time in activities like puzzles that call for insights because the act of struggling and then gaining understanding is so satisfying. We are built to seek insights.

But where do our insights come from? As I started collecting stories about insights, I didn't find any common strategy. I began my little project to survey some examples of insight in order to see what they had in common and to try to find some advice for pursuing the up arrow in the equation. However, as I compared the different stories, I got caught up in the mystery of what causes people to arrive at an insight in the first place. Each incident seemed different from the others. The stories of insights seemed to contain many clues, but I couldn't see how to make sense of them.

CHAPTER TWO

The Flash of Illumination

Almost a century ago, Graham Wallas, a cofounder of the London School of Economics, published the first modern account of insight. His 1926 book, *The Art of Thought*, contains a model that is still the most common explanation of how insight works. If you do any exploration into the field of insight, you can't go far without bumping into Wallas, who is the epitome of a British freethinking intellectual.

In contrast to my struggles to make sense of the incidents in my collection, Wallas succeeded in finding clear themes in the collection of insight stories he compiled. Perhaps I could learn from his example. His insight model might even answer questions about how the up arrow works and how to boost insights. It was time for me to get a history lesson.

Wallas was born in 1858 in Monkwearmouth, Sunderland, in the northeast corner of England. His father, a minister, gave his son a standard religious upbringing, but Wallas abandoned his Christian faith as a student at Oxford University from 1877 to 1881. He replaced it with a new faith, socialism. He joined the Fabian Society in 1884, shortly after it was founded by Sidney and Beatrice Webb. Fabians wanted to transform society by gradual, rather than revolutionary, means. Members included Bertrand Russell, Leonard and Virginia Woolf, H. G. Wells, and other luminaries. The society was named after the Roman general Fabius Maximus, nicknamed The Delayer, who avoided open battle with the

Carthaginian leader Hannibal, relying instead on continual pressure.

The Fabian Society was an advocate for causes such as minimum wage, slum clearance, a universal health care system, and a national education system. Several members of the Fabian Society helped form the British Labour Party in 1900.

Wallas quickly moved into the inner circle of the Fabian Society and formed a close connection with the Webbs. When Sidney and Beatrice Webb established the London School of Economics in 1895, they asked Wallas to be its first director. He turned down the offer but agreed to teach at the college. In her diary, Beatrice Webb remembered Wallas as a tall but slouching man, with pleasant features, driven by moral fervor rather than ambition. Although Wallas seemed a bit preachy to Beatrice Webb, he had a genius for teaching and inspired his disciples.

Among his various identities, Wallas was a psychologist. He believed that psychology could be used to improve society, especially to reduce some of the stresses created by the Industrial Revolution. He disagreed with the theory that people behave rationally and base their behavior on calculating the costs and benefits of each possible course of action. Wallas argued that politicians who want to get people to behave sensibly will need to study psychology.

In *The Art of Thought,* Wallas tried to apply concepts of psychology to show people how to think more effectively. Some contemporaries of Wallas—William James and Henri Bergson—had also speculated about the nature of insight, but neither provided an account as comprehensive and compelling as the one Wallas produced. He drew on his forty years of experience as a teacher and administrator; accounts that poets, scientists, philosophers, and others had published on their thought processes; and examples from students and friends.

The most lasting contribution of *The Art of Thought* is contained in a chapter called "Stages of Control," in which Wallas presents a four-stage model of insight: preparation, incubation, illumination, and verification.

During the *preparation* stage we investigate a problem, applying ourselves to an analysis that is hard, conscious, systematic, but fruitless.

Then we shift to the *incubation* stage, in which we stop consciously thinking about the problem and let our unconscious mind take over. Wallas quoted the German physicist Hermann von Helmholtz, who in 1891 at the end of his career offered some reflections on how this incubation stage feels. After working hard on a project, Helmholtz explained that "happy ideas come unexpectedly without effort, like an inspiration. So far as I am concerned, they have never come to me when my mind was fatigued, or when I was at my working table. They came particularly readily during the slow ascent of wooded hills on a sunny day."

Wallas advised his readers to take this incubation stage seriously. We should seek out mental relaxation and stop thinking about the problem. We should avoid anything that might interfere with the free working of the unconscious mind, such as reading serious materials. Wallas quoted the poet John Drinkwater about the way insights come about:

Haunting the lucidities of life
That are my daily beauty, move a theme
Beating along my undiscovered mind.

Next comes the *illumination* stage, when insight bursts forth with conciseness, suddenness, and immediate certainty. Wallas believed that the insight, the "happy idea," was the culmination of a train of unconscious associations. These associations had to mature outside of conscious scrutiny until they were ready to surface.

Wallas claimed that people could sometimes sense that an insight was brewing in their minds. The insight starts to make its appearance in fringe consciousness, giving people an intimation that the flash of illumination is nearby. At this point the insight might drift away and not evolve into consciousness. Or it might get interrupted by an intrusion that causes it to miscarry. That's why if people feel this intimation arising while reading, they often stop and gaze out into space, waiting for the insight to appear. Wallas warned of the danger of trying to put the insight into words too quickly, before it was fully formed.

Finally, during the *verification* stage we test whether the idea is valid. If the insight is about a topic such as mathematics, we may need to consciously work out the details during this final stage.

Wallas's four-stage model of insight is still the way most people explain how insight works. It's a very satisfying explanation that has a ring of plausibility—until we examine it more closely.

Wallas claimed that a preparation stage is necessary for insights to occur, but none of the five people discussed in Chapter One—the young cop, Martin Chalfie, Harry Markopolos, Michael Gottlieb, or I—spent any time preparing for an insight. Each insight came unexpectedly. Each was a surprise.

Now, in all five cases the protagonists drew on their background and expertise, but that isn't the same thing as deliberate preparation. The police officer had nothing to prepare as he sat in traffic. Chalfie would have had no reason to prepare for a biological marker. He wasn't looking to improve his methods. Likewise, Markopolos wasn't preparing to nail Bernie Madoff. He had to be goaded into examining Madoff's financial results. Gottlieb wasn't preparing to sound the alarm on the AIDS epidemic. No one knew about AIDS or anticipated how virulent it would be. And I wasn't preparing to reengineer my arrangement with my mechanic. The advice to begin by preparing wouldn't have helped any of us. We wouldn't have known what to prepare.

It is easy to confuse preparation with expertise. After we know what the insight is, we can see how the person gaining it acquired special kinds of knowledge. These previous interests and experiences prepared the person's mind to register the insight in ways that others missed. We can call this a *generally* prepared mind, a characteristic of Chalfie, Markopolos, and Gottlieb. None of the three would have gained his insight without years of special experience. A generally prepared mind is the same thing as having expertise. The young cop didn't have much experience, but he did have a mind-set to be alert for criminals. (And I didn't have any experience worth considering.)

Wallas, however, recommended that we have a *specifically* prepared mind by making deliberate preparations to solve a thorny problem. According to Wallas, when we're stuck and need to find an insight that will get us past an impasse, we should start with deliberate preparation. A few decades later this preparation activity was illustrated in one of the greatest achievements in science. James Watson and Francis Crick worked very hard to identify the structure of DNA (deoxyribonucleic acid) and eventually discovered that it was a double helix. If they hadn't gone through so much deliberate and specific preparation, they wouldn't have achieved their finding.

While the idea of deliberate preparation appeals to our work ethic—and is, of course, crucial for many types of work—it was not a factor in the insights attained by the cop, Chalfie, Markopolos, Gottlieb, or me. So I don't think deliberate preparation is necessary or even practical for many insights.

One flaw in Wallas's method is that his sample of cases was skewed. He only studied success stories. He didn't consider all the cases in which people prepared very hard but got nowhere. In the DNA example, researchers who were more respected at the time than Watson and Crick, such as Rosalind Franklin and Linus Pauling, were also working hard to discover the structure of DNA but didn't succeed. Deliberate and specific preparation doesn't guarantee success. Therefore, I didn't see how I could advise people to start with a preparation stage when so many insights are accidental and when specific preparation doesn't reliably lead to breakthroughs.

The incubation stage also doesn't fit most of the Chapter One examples. Gottlieb had time for his impressions to incubate between the different AIDS patients he examined. I had a few hours before having dinner with my wife and coming up with my car key scheme. None of the others had any time to incubate. The cop saw the driver ash his car and right away knew something was wrong. Chalfie heard about the green fluorescent protein and immediately saw the implications. Markopolos looked at the financial data and immediately sensed something was fishy.

Whereas preparation appeals to our work ethic, incubation appeals to our play ethic. It feels like a welcome time-out. Lots of people can recall insights that suddenly came to them while they were taking showers, but I doubt that organizations could increase the rate of insights in their workers by having them double the number of showers they take. The incubation stage doesn't seem necessary and often would be impossible.

Then there's the third stage of Wallas's model, the flash of illumination. Yes, we all had it, although Gottlieb arrived at it more slowly as he spotted the similarities among his patients. How does this flash of illumination work? Wallas describes it as the result of a train of unconscious associations.

That explanation seems too magical to be satisfying. This was the process I wanted to examine further, the mystery I started pursuing. What happens during this third stage? If I were going to tell people more about the up arrow, I'd need a better account of the flash of illumination.

What was I trying to explain, the illumination or the flash? Perhaps the "aha" experience, when everything finally snaps into place, marks the culmination of the insight process. Perhaps it isn't the insight itself.

We can use an analogy here: "aha" is to insight as orgasms are to conception. In both cases the experience is more noticeable than the achievement, but the experience doesn't guarantee the achievement, and the achievement can happen without the experience.

So what mystery was I trying to solve? At times, I felt like a bull charging forward at a swirling cape, hoping to make contact with a shadowy matador.

Wallas's four-stage model of insight isn't a good fit with any of the five people in Chapter One. That doesn't mean the model is wrong. Some aspects of it may be useful, but which ones?

Most cognitive scientists talk about insight as moving from an impasse state to a solution state. We hit an impasse, struggle for a while, and then find a way to overcome our mental block. This view is pretty much the same as Wallas's first three stages. But I don't see impasses with the cop, Chalfie, or any of the other cases in Chapter One. No one was stuck while trying to solve a problem.

If I wanted to describe how the up arrow works, I'd need at least a vague definition, some criteria, for what counts as an insight. I'd need something better than getting past an impasse.

AN UNEXPECTED SHIFT TO A BETTER STORY

The five people in Chapter One were shifting to a better story about how things work. These weren't entertainment stories. They were stories about what causes things to happen. The stories described the causes for past and present events (the young cop, Markopolos, Gottlieb) or ways to cause future outcomes (Chalfie as well as my car keys). These shifts weren't about making minor adjustments or adding more details. The shifts changed some of the core beliefs the five people initially held. During this transition some initial beliefs were abandoned or replaced. The shifts were *discontinuous discoveries*—unexpected transitions from a mediocre story to a better one.

Sometimes the shift landed immediately on the better story, as in the cop, Chalfie, and car keys examples. Other times the shift was toward the better story, but took a while to get there. Harry Markopolos's insight put him on the road to finding a better story about how Bernie Madoff was cheating. Similarly, Michael Gottlieb's detection of an ominous pattern of symptoms put him and the medical community on the road to identifying a new disease, AIDS, and the virus that caused it. Insights shift us toward a new story, a new set of beliefs that are more accurate, more comprehensive, and more useful.

Our insights transform us in several ways. They change how we understand, act, see, feel, and desire. *They change how we understand.* They transform our thinking; our new story gives us a different viewpoint. *They change how we act.* In some cases insights transform our abilities as well as our understanding; the Chalfie and car keys examples show how insights change our notions of what we can do. These shifts went beyond a richer story about how the world works. The new story was about how to make it work better, by using the green fluorescent protein in Chalfie's case and by giving keys to a car mechanic in my own example. *Insights transform how we see;* we look for different things in keeping with our new story. Gottlieb examined his new patients differently once he picked up the AIDS pattern. *Insights transform how we feel*—what excites us or makes us nervous. The two police officers got amped up when they suspected they'd found a car thief. Markopolos began with dread that Madoff was more skillful in financial investments than he was, then shifted to skepticism, and then to outrage that fueled his campaign to uncover Madoff's scam. Finally, *insights change our desires;* the new stories shift our goals, leading us to give up some ambitions and pursue others. Chalfie needed to find techniques for implanting the green fluorescent protein into the nerve cells of his worms. Markopolos needed to nail Bernie Madoff. Gottlieb needed to get to the bottom of a mysterious disease.

As a friend summarized all these transformations, "Insight is when it happens, everything that happens afterward is different." Hilary Mantel made the same observation in *Wolf Hall:* "Insight cannot be taken back. You cannot return to the moment you were in before."

So I had a working definition of insight—an unexpected shift to a better story—along with the ways insights transform us. I also had some ideas about what makes insights unique. Compared with routine problem solving, insights aren't conscious or deliberate. They come without warning. Our minds do their heavy lifting without our awareness. Watson and Crick labored to build a model of DNA, but the eventual insight—that it was a double helix—came as a surprise even to them.

Insights are unique in some other ways. When they do appear, they are coherent and unambiguous. They don't come as part of a set of possible answers. When we have the insight, we think, "Oh yes, that's it." We feel a sense of closure. This sense of closure produces a feeling of confidence in the insight. Wallas claimed that the flash of illumination results in a feeling of certainty. We aren't picking an idea that seems better than others. Instead, we're struck that this is the answer, the happy idea. We may even have an aesthetic reaction to the beauty of the insight. Watson and Crick felt that their double helix model was too beautiful not to be true. Chalfie may have had this experience when his new research agenda fell into place. Markopolos felt it when his puzzlement shifted into certainty that he had uncovered a clear case of fraud. Gottlieb had a grim sense of satisfaction when the new cases conformed to the pattern he had identified. The older police officer wanted to hug his partner.

Now I was ready to pursue the mystery of what produces the flash of illumination. Yet the more I learned, the more complexity and confusion I experienced. I wasn't sure if I would come up with anything useful. I didn't have a sense that I was getting warmer.

A NATURALISTIC INVESTIGATION

When I started my investigation into the flash of illumination, I experienced all the usual doubts I have at the beginning of a project, plus some new ones. I hadn't done any previous research on insight or even played around with the traditional insight tasks that cognitive scientists have used for almost a century. These puzzle tasks don't seem to have any possible solution until you find a way out of the impasse.

But perhaps that could be an advantage. I wanted to explore how insights work in more natural settings. I could then examine the flash of illumination without getting trapped by all the constraints surrounding laboratory-based experiments. The puzzle tasks used in the laboratory held little interest for me.

Twenty-five years earlier I had the same feelings when I set out to examine the way people made decisions. I'd never done a decision making experiment or even taken a course in decision making. I was trying to solve a different mystery at that time: how people can make life-and-death decisions under extreme time pressure and uncertainty. Instead of working in a laboratory, I chose to study how people actually make these kinds of decisions. You can't study that in a laboratory.

I studied firefighters because they have to make good decisions in stressful situations. I didn't give the firefighters any standard decision tasks. Instead, my fellow researchers and I interviewed the firefighters about their toughest challenges. We collected their best stories, and we probed those stories to learn more about the firefighters' strategies.

What we found surprised us. Our results didn't fit any of the existing models of decision making. The firefighters weren't trying to compare options to select the best. They didn't have time. Instead, they relied on their experience, the patterns they had acquired over decades, to quickly size up situations and recognize the option most likely to work.

The firefighters made rapid decisions by recognizing how the situations they encountered fit the patterns they had learned. The pattern-matching part of their decisions was fast and automatic. It was how they used their intuition to quickly identify an option that was likely to succeed. Then they evaluated their intuitions, not by comparing the option they recognized with others, but by consciously and deliberately imagining how it would fare if they carried it out. The firefighters relied on this strategy to make more than 80 percent of their tough decisions.

Other researchers have replicated our findings and showed that different specialists, such as military commanders and oil-rig managers, also rely on what we called the recognition-primed decision strategy for about 90 percent of their challenging decisions. These efforts helped to start the field of naturalistic decision making, which studies the way people think in natural settings, rather than how they are forced to think in laboratory experiments using artificial tasks.

Studying how people think in natural settings unnerves many research psychologists. They question whether naturalistic studies are real science. For them, science depends on testing ideas under tightly controlled conditions. Naturalistic decision making researchers like me concentrate on the early stages of the scientific method to provide the ideas that can later get tested. We try to observe and understand phenomena such as how people make tough decisions and how they form insights.

Traditional decision researchers hadn't come up with the recognitional strategy because their laboratory work on decision making typically studied novices trying to perform unfamiliar tasks. Their theories of decision making didn't take expertise into account. Recognitional decisions depend on decades of experience to build up hundreds and thousands of patterns.

And now, déjà vu. Almost all the studies of insight were taking place in laboratories and using artificial tasks. These studies were aimed at formulating and testing hypotheses. I saw a parallel between the field of insight and the field of decision making twenty-five years earlier, giving me some hope that a naturalistic study of insight might succeed. I didn't have any idea of how insight works, so I wasn't ready to test hypotheses. And that was the point. I could perform a naturalistic study of insight—explore the way people actually form insights—and if I were lucky, I might find some useful ideas in my stack of stories.

The work with firefighters might suggest that insights are the same as intuitions, but they're actually very different. Firefighters build up patterns that they apply in making rapid decisions in emergencies. *Intuition* is the use of patterns they've already learned, whereas *insight* is the discovery of new patterns.

Although insight differs from intuition, the firefighter study influenced my thinking about insights twenty-five years later. Firefighters often changed their beliefs about a complex fire as they learned more details, usually adding those specific details into their story. The most dramatic scenarios, however, contained surprises that forced the firefighters to rethink what was going on and replace erroneous beliefs. The firefighters shifted the story they told themselves about what was happening inside the burning structures.

Stories are a way we frame and organize the details of a situation. There are other types of frames besides stories, such as maps and even organizational wiring diagrams that show where people stand in a hierarchy. My work centers on stories because they are common ways we frame the events in the situations we encounter. These kinds of stories organize all kinds of details about a situation and depend on a few core beliefs we can call "anchors," because they are fairly stable and anchor the way we interpret the other details. In some later research I found that most of these stories build on only three or four anchors.

The cop's anchors were the expensive car and the driver's indifference to its upkeep. Chalfie's anchors were the transparency of the worms, the fact that the green fluorescent protein could be made to glow, and the belief that proteins like this could be implanted in neurons inside his worms. Markopolos's anchors were that investment funds that never lost money in the stock market were usually fraudulent, and that the method Madoff claimed to be using was too conservative to generate large profits. Gottlieb's anchors for the story he was forming were that he had encountered a disease that targeted gay men, that it devastated part of their immune systems, and that it left the victims vulnerable to all kinds of infections.

Anchors can change as a person gets more information. Chalfie's story got richer as he learned how to manufacture the green fluorescent protein and how to insert it into a variety of cells. Markopolos's story evolved as he learned more about the kinds of fraud Madoff was committing. Gottlieb came to learn that AIDS did not afflict only gay men.

Actually, I wasn't thinking of any of these ideas when I began investigating the flash of illumination. I wasn't thinking of firefighters or anchors or the ways we frame and organize events. It is only now, with hindsight, that I appreciate how my earlier studies enabled me to form a richer explanation of how insights work. At the time what the earlier projects with firefighters and others gave me was hope that I might be able to make a discovery here, even though I hadn't done any experiments with insight puzzle problems, or perhaps *because* I had not done any of these kinds of studies. I was ready to examine cases of actual insights, starting with the clippings in my stack.

ARCHAEOLOGICAL TRENCHES

When field archaeologists want to explore the structures inside an unnatural hill, they don't excavate the entire hill. That level of effort would be too costly and take too long. Instead, they dig a trench that gives them a snapshot of what's inside the hill. It helps them assess whether the site has walls or housing materials or any other significant structures. The archaeologists may judge that the hill is not worth further investigation, or they may decide that the site has enough potential for a fuller excavation.

My research plan was like an archaeological trench. I didn't conduct the kind of laboratory experiment that would constitute a traditional study of insights. Instead, I searched for insights of people who made an unexpected shift in their beliefs. I collected a set of incidents—120 examples, including the small number in my original stack. Once I had assembled them, I could review them to see if I could find any themes. With luck, somewhere in those stories I would run across some clues to what was behind the flash of illumination.

It took me more than half a year to compile my sample of 120 cases in my spare time. I didn't do much interviewing because I found so many examples in the media. I relied heavily on books, especially those describing innovations and discoveries. I also used newspaper and magazine articles. I drew on interviews I'd conducted over the previous thirty years and selected those in which people unexpectedly made radical shifts in their stories and beliefs about how things work.

The books on innovation yielded a harvest of examples from science, inventions, business, and management. I searched through transcripts of interviews I had conducted with military decision makers. Books and articles about financial investment in the wake of the 2007–2008 financial collapse yielded many more examples, as did stories about medical discoveries. I trolled for examples in my discussions with professional colleagues and even with family members. The 120 stories were the only data I had. I didn't anticipate where these stories would lead or how they might fuse to create their own story. That came later as I was finishing the project. Perhaps as you read the insight stories in these early chapters, you will form a better picture than I had.

My collection came together in a somewhat haphazard way. When I felt I could learn from an incident, I studied it further. Researchers in the future may come up with more systematic ways to assemble incidents, but at this point, doing an initial naturalistic study of insight, I just needed to gather observations. I needed to start digging.

Most of the 120 stories are about successes, the same criticism I made of Graham Wallas. An investigator can pull together a bunch of success stories and conclude that a certain strategy, such as careful preparation, invariably results in insights. We won't see the failures, the cases in which preparation didn't pay off. To counter this weakness, late in my project I selected a small set of cases that each had a built-in control. These cases described not only the person who had the insight, but also another person who had access to the same data yet didn't achieve the insight. That's one of the criteria for insights: that others with the same information somehow miss the discovery. This control sample highlighted the strategies and traits that make a difference between success and failure. I'll describe this comparison later in the book in Part II.

As I assembled the 120 examples of insights, I summarized each one in a short account, three pages or less, sometimes only a half page. I divided each account into three sections: background on the incident, the critical events that led to the insight, and the nature of the insight.

Next, I coded the incidents to try to uncover patterns. I defined a set of coding categories. The list started small, but I added new features that I wanted to learn more about, and the set eventually grew to fourteen categories as the project went on. This was a trench, not a controlled experiment. Its purpose was to explore insights.

The categories were issues like whether the person struggled with an impasse (yes or no), whether the person had time to incubate (yes or no), whether the insight was sudden or gradual, and so forth. Once I had defined the categories, my research assistant, Andrea Jarosz, and I independently coded each incident using each of the fourteen categories. Our ratings had a 78 percent rate of agreement, which wasn't a bad start. Then we discussed the cases on which we disagreed and adjusted our criteria to reduce confusion. I also added more details to some of the stories. By the end our interrater agreement was 98 percent.

Eventually I was able to sort these 120 cases into five different strategies for gaining insights: connections, coincidences, curiosities, contradictions, and creative desperation. Did the incident rely on a person making a connection? Did the person notice a coincidence as a trigger for the insight? Was the insight triggered by some curiosity—an odd fact or event? Did it depend on seeing a contradiction? Or was the person stuck, desperately seeking some way out of an impasse?

All the 120 cases fit one of these strategies. Most relied on more than one of the five strategies. But which of the five was the best strategy, the "real" one that explained insights? Or should all five be combined into one general strategy? Neither of these approaches seemed promising to me as I struggled to find a pattern in the data, but you can judge that for yourself. The next four chapters will describe the five candidate strategies. We'll start with the prime suspect.

CHAPTER THREE

Connections

Martin Chalfie is a perfect example of the experience most people have of "connecting the dots" and solving a problem by being exposed to more ideas. Like Chalfie, we get a new piece of information that combines with other information we already have, and, presto, we make a discovery. Let's see how it happens by looking at the routes taken by the admirals who noticed that Pearl Harbor was vulnerable to attack, a prominent child psychologist who made important discoveries about the empathy of babies, and Charles Darwin, who formulated a theory of evolution.

THE BATTLE OF TARANTO

Few people have heard about the Battle of Taranto, which took place November 11–12, 1940, even though it forever changed the way military leaders think about combat at sea. The battle showed that battleships were vulnerable to airplanes launched from carriers.

The battle took place at the beginning of World War II before the United States entered the conflict. France had surrendered on June 25, 1940, leaving the British Empire alone to combat Germany, Italy, and Japan. In the Mediterranean Sea, the Italian fleet restricted British efforts to resupply British forces in Egypt. The British devised Operation Judgement to attack the First Squadron of the Italian Navy, which was comfortably sheltered in the Bay of Taranto on the inside of the heel in the southeast of Italy. To maintain secrecy, no written records were made of the planning that went into this mission, which explains why it has received so little attention.

The British used a single, brand-new aircraft carrier, the HMS *Illustrious,* to mount the first all-carrier air attack in history. The first airplanes took off at 9:00 in the evening on November 11 and reached their target at 10:58. In less than an hour the British airplanes launched their torpedoes and put half the Italian fleet out of commission for six months.

The British carried out the attack with vintage biplanes, twenty-four British Swordfish bombers. Up to that time, experts believed that torpedo attacks could be made only when the water depth was at least 30 meters (98 feet), the depth heavy torpedoes sank after being dropped from an airplane. And Taranto Bay was only 40 feet deep. The Italians believed their fleet was safe from airborne torpedoes.

The British, however, had devised a way to use a wire to pull up the nose of the torpedo so that it did a belly flop rather than a nosedive. They also added wooden fins to the torpedoes so that they wouldn't dive so deeply.

When news of the British success at Taranto got out, the Japanese Admiral Isoroku Yamamoto saw the attack's implications. He had the insight that the American naval fleet "safely" anchored at Pearl Harbor might also be a sitting duck, vulnerable to an unexpected air attack. Yamamoto first wrote down his ideas in January 1941 and continued refining them until they became the blueprint for the Japanese attack on Pearl Harbor on December 7, 1941.

Somewhat ironically, Yamamoto did not approve of Japan's decision to go to war with the United States because he was pretty sure Japan would lose such a war. In a letter whose contents were leaked, he wrote, "Should hostilities once break out between Japan and the United States, it would not be enough that we take Guam and the Philippines, nor even Hawaii and San Francisco. To make victory certain, we would have to march into Washington and dictate the terms of peace in the White House. I wonder if our politicians (who speak so lightly of a Japanese-American war) have confidence as to the final outcome and are prepared to make the necessary sacrifices."

Yamamoto had studied at Harvard University for a few years and was fluent in English. He opposed the Japanese invasion of Manchuria, its invasion of China, and its pact with Nazi Germany, as well as its preparation for war with the United States. His opposition to these military activities made him a target for assassination by Japanese militarists. To protect Yamamoto's life, the naval minister appointed him the commander in chief of the combined forces and sent him off to sea, where he'd be safe.

In contrast to Yamamoto, the Japanese army leaders were confident that Japan would prevail and were eager to brush the United States aside to pursue their plans to dominate Southeast Asia. These leaders had too much political power to be stopped. Yamamoto knew that with war inevitable, Japan couldn't outlast the United States, but perhaps starting the war with a knockout blow that devastated American naval power would be a way for Japan to triumph.

Yamamoto delivered that blow at Pearl Harbor with his surprise attack, using 353 airplanes from six aircraft carriers. The Japanese forces hit all eight battleships at Pearl Harbor, sinking four of them and badly damaging the others. The Japanese also sank or damaged many other ships, including three cruisers and three destroyers; wrecked 188 U.S. airplanes; and killed 2,402 Americans.

Admiral Harold Stark, the chief of naval operations (CNO) for the United States, saw the implications of Taranto even more quickly than Yamamoto. Less than two weeks after the Battle of Taranto, Stark signed a memo, dated November 22, 1940, stating, "By far the most profitable object of a sudden attack in Hawaiian waters would be the Fleet units based in that area." The memo observed that it might be desirable "to place torpedo nets within the harbor itself." Later, on January 24, 1941, Stark sent a letter to his boss, Frank Knox, the secretary of the navy, stating, "If war eventuates with Japan, it is believed easily possible that hostilities would be initiated by a surprise attack upon the Fleet or the Naval Base at Pearl Harbor."

Two admirals in two opposing navies quickly grasped the significance of the British attack. They both had the same insight. The Japanese turned that insight into a battle plan for a surprise attack. The Americans lost track of the warning issued by Admiral Stark, the head of the entire U.S. Navy.

Why didn't Stark follow up on the threat against the American fleet? He tried to. He issued the warning. He suggested the torpedo nets. But it is hard to maintain vigilance for a long period of time. Daily emergencies create constant distractions. Plus, the energy put into self-defense gets in the way of training. Besides, torpedo nets interfere with ships moving in and out of the harbor. And how plausible was this threat that Stark warned about? Pearl Harbor was so shallow that the admiral in charge of its defense didn't believe the Japanese could attack the American ships with torpedoes dropped from airplanes. The torpedoes would get stuck in the mud. Apparently, the navy officials who knew that the Japanese had developed shallow-water torpedoes never bothered to pass this information on to the Pearl Harbor commanders. Therefore, no torpedo nets were put into place.

Insights don't count for much if we can't translate them into action, and despite Stark's position as CNO, his subordinates deflected and ignored his insight. President Franklin D. Roosevelt removed Stark as CNO in March 1942 and reassigned him to London, an unofficial way of blaming him for the disaster.

Consider the insight Yamamoto and Stark shared. They both anticipated the coming war between their countries. They both knew that the U.S. Navy had superior forces. And then this news about Taranto. Mapping the Battle of Taranto onto the Pacific, Yamamoto and Stark recognized that it was an analogue for what the Japanese might be able to do to the U.S. fleet in Pearl Harbor. Yamamoto and Stark both made the same connection.

This connection process seems like the same one Martin Chalfie followed in grasping the significance of the genes that glowed with green light. All three men, Yamamoto, Stark, and Chalfie, received new and unexpected information and grasped its implications. They quickly connected the new data to their situation to discover insights about actions they could take.

BROCCOLI AND GOLDFISH

Alison Gopnik formed a connection insight from a remark made by her two-year-old son. Gopnik is a highly respected developmental psychologist at the University of California at Berkeley. She comes from an accomplished family: one younger brother is Blake Gopnik, the *Newsweek* art critic, and another is Adam Gopnik, staff writer for the *New Yorker* magazine. One of her specialties, besides inspiring younger brothers, is understanding how babies think.

Gopnik believed that babies are smarter than people realize. She suspected that in some ways babies could take the perspectives of other people. In psychology-speak, this skill is described as theory of mind—an appreciation that other people have their own minds and preferences. She believed that children make sense of the world in the same way scientists do, forming stories/theories about how causes work. Gopnik dislikes the notion that babies are just defective, unformed grownups. She sees babies as different and, in some ways, more alert to their surroundings than grownups and more active learners than scientists. They are designed to learn about the world around them, not just their own specialty field. Grownups often narrow our attention to what matters most to us. Babies aren't as good at attending to just one thing but seem to be more alert to anything unexpected. Gopnik's 2009 book, *The Philosophical Baby,* describes the wonderful capabilities she has been able to demonstrate, such as the capacity for empathy.

Developmental psychologists had believed that children didn't develop the ability to take another perspective until they were about seven years old. Gopnik thought decentering—taking another person's perspective—happened much earlier. But she didn't know how to demonstrate that babies have this skill.

Then came a fateful day when Gopnik made a fancy dessert, pineapple in kirsch, for a dinner party she was hosting. Her two-year-old son had a taste of it and made a terrible face. "Then for weeks afterward, completely out of the blue, he would turn to me and say, 'Mommy, pineapple is yummy for you, but it's yucky for me.'"

Gopnik turned her son's comment into an experiment, the broccoli and goldfish paradigm. She gave fourteen- and eighteen-month-old babies bowls of raw broccoli and Pepperidge Farm Goldfish Crackers and had them try some of each. They all liked the crackers more than the broccoli. Then Gopnik's co-researcher, Betty Repacholi, took a bit of food from each bowl and made either a disgusted face or a happy face. Half the time Repacholi made a disgusted face with the goldfish and a happy face with the broccoli, and the other half of the time it was reversed.

Repacholi would then put her hand out to the baby and ask for some food. The eighteen-month-old babies would offer the type of food that Repacholi liked, even if it was the broccoli, not the goldfish that the baby liked. And in case you suspect the babies were hoarding their goldfish, if Repacholi expressed a liking for goldfish, that's what the babies generously offered. The fourteen-month-old babies, however, just got confused when Repacholi went into raptures over the broccoli, and then they offered her the goldfish.

Using the broccoli and goldfish task, Repacholi and Gopnik demonstrated that eighteen-month-old babies were capable of empathy, but not fourteen-month-olds. The eighteen-month-olds didn't respond egocentrically, giving Repacholi the food (goldfish) that they liked, nor did they respond selfishly, hoarding the goldfish. Instead they offered Repacholi the food she seemed to like the most. Experiments like this opened the way for psychologists to make discoveries about the way babies think.

Gopnik's insight follows the same trajectory we've seen before. She encountered a new and unexpected idea and quickly saw how it linked to the work she was doing. Like Yamamoto, she arrived at a new understanding that expanded the actions she could potentially take. Insights like these show us more effective ways to get things done and even expand our goals. For Yamamoto, the insight gave him an idea for how to cripple the American fleet. Stark discovered he needed to take additional measures to protect that fleet, but he tragically failed to get his subordinates to heed his warning. Gopnik designed a new method for testing infants.

THE MOTHER OF ALL SCIENTIFIC INSIGHTS

Charles Darwin followed the connection path in formulating his theory of evolution driven by natural selection. Darwin was only twenty-two years old when he set out on the voyage of HMS *Beagle* on its mission to chart the South American coast. A recent graduate of Cambridge, Darwin expected to become an Anglican parson. He wasn't following any family tradition. His father and grandfather were freethinkers, and as a child he attended a Unitarian chapel with his mother. Perhaps he was influenced by the Anglican boarding school he was sent to following his mother's death when Darwin was nine years old.

Darwin was indifferent to his classes when he attended the University of Edinburgh Medical School, so his father switched him over to Cambridge. The one topic that did capture his interest was natural history. He made some findings about beetles and became a follower of John Henslow, a leading botany professor.

Before settling down to a parson's life, Darwin decided to take a few years to pursue his long-standing interest in natural history. Henslow nominated Darwin for an unpaid position as a naturalist aboard the HMS *Beagle*. The captain of the *Beagle* was happy to have Darwin as a companion for the voyage, and Darwin was grateful for the opportunity.

The expedition lasted for five years, from 1831 to 1836. Darwin spent his time making observations about geography and about different animal species. It was a time of wonders for him as he encountered all manner of odd phenomena. At the ship's first stop in the Atlantic in Cape Verde, Darwin found seashells high up on a cliff. In Patagonia he discovered fossil bones belonging to huge mammals that were extinct. In the Galapagos Islands Darwin found mockingbirds that differed from one island to the next.

Upon his return to England, Darwin wondered what drove the variations in species he had seen on his voyage. He knew that farmers and pigeon breeders could deliberately cultivate favorable variations. Why would that occur in nature?

Then in September 1838, Darwin read *An Essay on the Principle of Population* by the Reverend Thomas Robert Malthus on population growth and the competition for resources. Malthus's essay, written forty years earlier, claimed that populations grow until they exceed the food supply and then the members of the population compete with each other. Darwin immediately saw how this could explain the variations in species he had observed. In the competition for resources, any random variation that created an advantage would be selected and others would lose out. The species members with the competitive advantage would be more likely to survive and breed and transfer their traits to offspring. Unfavorable variations would be lost. Darwin saw that nature was doing automatically what farmers had been doing deliberately—selecting the best traits and favoring their continuation.

The result was Darwin's theory of natural selection relying on blind variation and selective retention. Blind variation refers to the natural variability of physical features such as size and coloration in individual members of a species. Selective retention means that if a variation increases survival value, then the individual possessing that variation would be more likely to survive and pass the trait on to its descendants. (In 1871 Darwin added sexual competition and selection to the mix.) Reading Malthus gave Darwin the missing piece that fit all his observations together.

At roughly the same time, Alfred Russel Wallace, also a British naturalist, was doing fieldwork in the Malay Archipelago and came up with the same theory of evolution. Wallace had been galvanized by his own reading of Malthus.

Independently, Darwin and Wallace were wondering how species change. They were primed to connect their ideas to Malthus's notions of competition for resources. They both connected a set of ideas in ways that others did not. And they both had firsthand experience as naturalists. They had seen variations between species but also individual differences within a species. They both could appreciate how individual differences provided a platform for blind variation and selective retention to work. Lots of people read Malthus's book; Darwin read the sixth edition. But only Darwin and Wallace saw how to apply Malthus's ideas to the evolution of species.

All these cases (Yamamoto, Stark, Chalfie, Gopnik, Darwin, and Wallace) follow a connection strategy for making insights. The person making the discovery gets some new piece of information and sees how it combines with other information to form a new idea. Sometimes the person sees a new way to combine different kinds of information even if none of them are new.

THE LEADING CANDIDATE

The connection strategy seemed like the answer to the mystery I was pursuing. It accounted for 82 percent of the cases in my sample, 98 out of the 120 cases. It fit nicely with Graham Wallas's idea of an unconscious stream of associations that eventually burst into awareness. The strategy offers a clear image of insights as connecting the dots. And it suggests that we can increase insights by exposing ourselves to lots of different ideas that might help us form new connections. So I now had a coherent answer to give people who wanted to know more about the up arrow in the equation I showed them and wanted to know what they could do to have more insights.

Still, I wasn't ready to conclude that the connection strategy explained the flash of illumination. In fact, I had some reservations about this connection strategy. It suggests that insights are just a matter of connecting the dots, and I dislike the "connecting the dots" metaphor. The Yamamoto, Stark, Chalfie, Gopnik, and Darwin examples come across as connecting the dots because I haven't told you about all the non-dots. The non-dots would have slowed the stories down, but they are an important part of each insight—determining which dots to connect. Anyone can connect the dots if we remove the non-dots and clarify the ambiguous dots and group the dots that look different but are really the same. If all that is left is a set of unique and valid dots, then the job of connecting the dots becomes much easier. If we leave in all the ambiguities, then the connection gets trickier. "Connecting the dots" trivializes the business of making sense of events and arriving at insights.

Look at some of the non-dots in connecting the Battle of Taranto to Pearl Harbor. There must have been many non-dots—irrelevant messages—so let's just examine the ones that are "anti-dots"—data about the *differences* between Taranto and Pearl Harbor that could have disconnected the two harbors and blocked the insight gained by Yamamoto and Stark.

Start with the problem of getting to the target. For Taranto, the British warships were already in the Mediterranean. To attack Pearl Harbor, the Japanese would need a large convoy, at least six aircraft carriers instead of one. This convoy would have to travel for 4,000 miles without being detected. To travel that far, the aircraft carriers and the other ships in the convoy (whose size reached thirty ships in the actual Pearl Harbor assault) would need to be refueled. The British didn't have to worry about refueling in the Battle of Taranto, but at that time the Japanese didn't have the refueling capability they would need for the 4,000-mile journey to Pearl Harbor.

Next we add the problem of bad weather. The British didn't worry about weather at Taranto—the Swordfish torpedo bombers only had to fly for two hours to reach the Italian fleet. In contrast, the Japanese fleet would have to sail for a week and a half before reaching Pearl Harbor. What if the fleet arrived in the middle of a Pacific storm and couldn't launch a single airplane? Even today's meteorologists don't have that kind of long-range forecasting accuracy.

Once the Japanese fleet reached Pearl Harbor, it would have to contend with tougher defenses than the British faced at Taranto. The United States had its own aircraft carriers, which posed a severe threat in case they detected the Japanese approach. The British attacked the Italian fleet at night, but the Japanese lacked precision instruments and would have to mount a daytime attack. After Taranto, surely the United States would be flying reconnaissance missions around Pearl Harbor. (Actually, the United States ceased its aerial reconnaissance because those missions interfered with training.) And after Taranto, the U.S. Navy was likely to be using torpedo nets, blunting the primary Japanese weapon. (The U.S. commanders, as noted earlier, chose not to deploy torpedo nets, believing that their force was protected by the shallowness of the water at Pearl Harbor.)

That's a final anti-dot. At the time of Taranto, the Japanese couldn't successfully launch airborne torpedoes in such shallow waters. The Japanese had to work hard to get their torpedoes to work at Pearl Harbor.

The connection between Taranto and Pearl Harbor isn't so obvious when we toss in non-dots and anti-dots. We no longer look at Taranto and conclude how easy it would be for the Japanese to do the same thing at Pearl Harbor. By exploring the anti-dots, we see why Yamamoto and Stark deserve so much credit for not getting distracted by irrelevant details. They recognized the fundamental lessons of Taranto: that naval fleets were now vulnerable to air attacks and that concentrating battleships and cruisers in one place, such as Taranto and Pearl Harbor, makes them an inviting target. They become the anvil, not the hammer.

Many of Admiral Stark's subordinates were so fixated on the idea that the navy battleships were *protecting* Pearl Harbor that they had trouble seeing the ships themselves as a Japanese target. The connection strategy is more than just connecting the dots. It involves changing the way we think. We'll see how some of the other insight strategies—contradictions and creative desperation—shake up our thinking as much as or even more than the connection strategy.

That statistic I tossed out earlier—that connections were involved in 98 out of the 120 cases—is misleading. The other insight strategies also turned up a lot when I coded the data. I coded each incident for all five insight strategies: connections, coincidences, curiosities, contradictions, and creative desperation. Most of the insights were a blend. They depended on more than one of the strategies. Thus, of the 98 cases in which I found the connection strategy, only 45 used the connection strategy alone. The other 53 depended on making a connection, plus one or more of the other strategies. That's why we shouldn't be too quick to conclude that insights are simply about making connections. In a mystery story, you don't stop your investigation with the first likely suspect you encounter.

CHAPTER FOUR

Coincidences and Curiosities

When we notice a coincidence, we may not be sure what to make of it. Is it just an accident, or is some deeper pattern at work? Observing a coincidence means that we've spotted some events that seem related to each other even though they don't seem to have any obvious causal link.

Coincidences are chance concurrences that should be ignored except that every so often they provide us with an early warning about a new pattern. When Michael Gottlieb encountered a second and then a third patient with a compromised immune system, he might have dismissed it. Instead he became suspicious. Something was going on, something he didn't understand, and he needed to monitor it more carefully. Gottlieb didn't believe his patients had anything to do with each other. But the coincidence in their symptoms seemed important. And the men were all gay. Was there any significance in that? Quickly, the coincidence turned into a pattern, the deadly pattern of AIDS.

We tend to notice coincidences, associations we don't fully understand, based on relationships we can't articulate. People who can pick up on trends, spot patterns, wonder about irregularities, and notice coincidences are an important resource. They may often be wrong, so we shouldn't automatically believe them even if they feel very confident. Nevertheless, they should be listened to, rather than ridiculed, because they just might be on to something.

At first, I was unsure what to make of the incidents in my research sample that were triggered by coincidences. They didn't immediately result in a better story. Instead they shook people loose from the initial stories they had held. I decided to count these incidents as insights because they sent people on a road that eventually led to a better story.

THE GREATEST ASTRONOMICAL DISCOVERY OF THE TWENTIETH CENTURY

Consider the case of Jocelyn Bell Burnell, an Irish astrophysicist who turned a coincidence into a Nobel Prize. Her father was the architect for the Armagh Observatory southwest of Belfast, and she used to accompany him during inspection visits starting when she was seven or eight years old. She remembers being sent to crawl around in the rafters searching for leaks in the roof.

As a teenager she became interested in astronomy, but she also liked a good night's sleep and couldn't imagine going into a field where she'd have to stay up all night to collect data. Then she found out about radio astronomy and X-ray astronomy, which allowed researchers to collect data during the day. So she'd be able to collect data without being sleep deprived.

When Burnell was a graduate student in astronomy at Cambridge University in the late 1960s, she built her own radio telescope (a device that evolved from World War II radar), using 120 miles of cable in a four-acre site. The data came out on pen recorders making squiggly lines on moving chart paper—like a lie detector, but for stars instead of criminals.

Burnell was twenty-three years old when she started her thesis project in 1968 to find mysterious objects called quasars, very distant objects that emitted a lot of energy.

In analyzing the data, she had to distinguish between the squiggles caused by quasars and those caused by other sources of radio interference. And then she noted a different burst activity: "There was another bit of squiggle that didn't make sense. It took up about a quarter-inch out of 400 feet of chart paper. I think the first few times I saw it, I noted it with a question mark. But your brain remembers things you don't realize it remembers. By about the fourth or fifth time I came across this signal, my brain said, 'You've seen something like this before.'"

To spread this new signal out and inspect it, she had to get more than a quarter inch of recording. She had to run the paper at high speed for five minutes at the appropriate time of day and then back at normal speed. She couldn't leave the machine at high speed throughout the day because doing so would consume too much paper.

She did that for a month and captured a detailed trace of this new type of signal. To her surprise, the burst of activity looked like a period of unusually regular pulses. She wondered what caused it and worried that there was something wrong with the equipment. But the maintenance technicians checked, and the activity wasn't crossed wires and it wasn't interference and it wasn't a satellite or a signal bouncing off a corrugated steel building. So what was it?

The signal pulsed very regularly, once every 1.339 seconds. Later, she was looking at a completely different part of the sky and again saw something strange. Again she came into the observatory (this time at 2:00 A.M.), switched on the high speed for the paper, "and in came *blip, blip blip,* this time one and a quarter seconds apart, in a different part of the sky. That was great. That was the sweet moment. That was eureka." So the first set of odd squiggles wasn't an anomaly. Soon after, she found a third example, and then a fourth.

So as a graduate student, Burnell discovered pulsars, which we now know are rapidly spinning neutron stars. They are the leftovers of supernova explosions and are massive but very small. They weigh as much as the sun but are only a dozen miles wide.

Soon after she and her advisor announced her finding, radio astronomers around the world were redirecting their equipment to search for pulsars. It became a bit of a craze, like a gold rush. The publication of this finding listed Anthony Hewish, Burnell's advisor, as first author. Hewish received a Nobel Prize in 1974, primarily for the discovery of pulsars.

The Nobel Committee came under some criticism for its omission of

Burnell, who did receive many other academic awards during her academic career. Some called it the "No-Bell Prize," because of the omission of Jocelyn Bell (her maiden name). One prominent astrophysicist told her in public, "Miss Bell, you have made the greatest astronomical discovery of the twentieth century." Burnell was diplomatic about the matter, stating that Nobel Prizes should be awarded to research students only in exceptional cases and her pulsar work was not one of those cases.

Burnell's insight began when she noticed a coincidence, the unusual squiggles on the pen-and-ink recorder. She remembered seeing similar bursts of activity from the same location. Then she found another coincidence when she picked up the same type of signal in another part of the sky. Rather than dismissing these signals as "mere coincidence," she homed in on them.

Burnell's example resembles Michael Gottlieb's detection of unusual symptoms shown by a small set of patients. Nobody knew about AIDS at the time. Gottlieb couldn't have had the insight that his patients had AIDS any more than Burnell could have known that she was looking at pulsars. Burnell and Gottlieb just felt that they might discover something useful by investigating the new coincidence they noticed.

Spotting a coincidence is like a hunter picking up a trail. Coincidences guide the way people search for evidence. When Burnell got interested in those mysterious bursts, she rescheduled her life, interfering with her much desired sleep, to get to the observatory and slow down the paper speed at the times when she thought she might capture the bursts in more detail. Coincidences change our understanding, change what we notice, change what excites us, and set us on the path to making a discovery.

Coincidences can also change our actions. One way they do this is by giving us an idea of what we need to alter to break a pattern we don't like, as shown in the game plan the Denver Broncos football team devised to thwart LeRoy Butler in the Super Bowl.

DEFENDING THE DEFENDER

Super Bowl XXXII, played on January 25, 1998, pitted the Denver Broncos against the Green Bay Packers. During Denver's preparations for the game, the team's coaches reviewed films from Packers games earlier in the season. The coaches noticed again and again that one of the Green Bay defensive backs, LeRoy Butler, was getting in on the action even when he should have been elsewhere on the field. He was an unblocked phantom who would appear out of nowhere and ruin the plays of his adversaries. The Denver coaches never quite grasped how Butler was able to do this. But it happened too often to be ignored.

Accordingly, the Denver coaches set a high priority to hit and deflect Butler. They created schemes so that different players would block him. Sometimes it would be a tackle, sometimes a fullback, sometimes a wide receiver. As a result, Terrell Davis, the Broncos' running back, was able to gain 157 yards with three touchdowns, winning the most valuable player award. Butler still led Green Bay with nine tackles, but he rarely got into the backfield to disrupt Denver's plays. The Denver coaches forged a winning game plan using the coincidence they detected: that LeRoy Butler usually seemed to be in the right place to disrupt a play.

Coincidence insights are different from connection insights. In the case of connection insights, such as the Chalfie and Yamamoto and Gopnik examples, the new pieces of information provided important details. The details count. If Chalfie had gone to a different lunch seminar, forget about the Nobel Prize. In contrast, what matters for coincidence insights is the repetition. If Gottlieb's first patient had been a no-show, the second and third and fourth patients would have been enough to illustrate the pattern, the common symptoms of AIDS. The details of any one patient didn't matter.

Coincidences were a second insight strategy in my sample. They cropped up in only 12 of my cases, just a tenth of the sample, but they illustrated another way that insights can form.

CURIOSITIES

Late in my project a third insight strategy emerged: curiosities. Some insights were sparked by a single event or observation that provoked the reaction "What's going on here?" These curiosity-driven investigations often led to impressive discoveries. One well-known example is the way Alexander Fleming stumbled upon the disease-fighting properties of penicillin. In 1928, Fleming noticed a halo of destruction affecting a bacterium, *Staphylococcus*. Fleming had been growing colonies of *Staphylococcus* in petri dishes, and he put them aside for the month of August while he took his family on vacation. When he returned, he found that one of the petri dishes had gotten contaminated with a mold. Oddly, all the *Staphylococcus* bacteria in the vicinity of the mold had gotten destroyed, whereas the pathogen colonies that weren't in contact with the mold were growing normally. Fleming wasn't expecting anything out of the ordinary, but when he spotted the unusual pattern in one of the petri dishes, he said, "That's funny." Fleming cultured the mold and found that it contained an infection-fighting substance—Fleming originally called it "mould juice"—that killed *Staphylococcus* and other bacteria. Further investigation led to the discovery of penicillin, the world's first antibiotic.

Curiosities provoke people to investigate further, just as coincidences do. The initial "What's going on here?" reaction doesn't contain the insight, but it starts the person on the road to gaining the insight. Curiosities differ from coincidences in one way: They are sparked by a single event or observation rather than by the repetition of a pattern.

Wilhelm Roentgen's discovery of X-rays in 1885 came about through a curiosity. He was investigating cathode rays. He used a cardboard covering to prevent any light from escaping his apparatus. Then he noticed that a barium platinocyanide screen across the room was glowing when his apparatus discharged cathode rays despite the cardboard covering. That was odd. So he stopped his investigation to look more carefully at what this was all about. After several weeks Roentgen satisfied himself that the effect was not due to the cathode rays but to some new form of light. At the time, physicists appreciated a number of forms of radiation: visible, infrared, and ultraviolet. So X-rays could have been added to the list. But that didn't happen. X-rays were greeted with disbelief. Lord Kelvin labeled them an elaborate hoax. One reason for the resistance was that a number of people used cathode ray equipment. If X-rays existed, surely others should have noticed them. (One researcher had noticed a glow but never accounted for it.) The skeptics eventually came around, and in 1901 Roentgen received the very first Nobel Prize in Physics.

Transistors were also discovered through curiosity. In the early 1940s Russell Ohl, an engineer at Bell Labs, tried using silicon to get better radio reception. One day something strange happened when he inadvertently used a piece of silicon that had a crack in it. When the silicon was exposed to light, the current flowing between the two sides of the crack jumped significantly. That aroused Ohl's curiosity. He found that the crack was a dividing line between two impurities in the silicon. Ohl performed additional research showing that the impurities changed how much each section resisted electrical flow. Ohl's discovery led to transistors and to all forms of diodes. He later built on his diode research to create the first silicon solar cells.

My sample had only 9 cases of curiosities, fewer even than coincidences. Curiosities are unlikely to get us in trouble. If we examine a curiosity that doesn't lead anywhere, we've just wasted some time. In contrast, coincidences can mislead us.

THE DANGER OF COINCIDENCE

We are all attuned to coincidences. We are sensitive to associations. Sometimes we are too sensitive and see connections that aren't real. That's why we are skeptical about "mere coincidence." Insight cannot depend on simply noticing coincidences. Burnell, Gottlieb, and the Denver Broncos coaches weren't pursuing every association, every coincidence they noticed. Their success depended on their ability to identify coincidences that might have important implications even if they didn't yet know what those implications might be.

All the coincidence stories in this chapter show people with the background and the expertise to judge that a coincidence might be important. Their sense of what is typical enabled them to wonder about something atypical that might have meaningful consequences.

We are association machines, forever noticing coincidences even when they are spurious. And many coincidences are spurious. This chapter has described cases in which coincidence paid off in insight. The chapter hasn't presented cases in which the coincidence turned out to be a "mere coincidence" that didn't pan out.

Therefore, we are taught that we need to test coincidences before giving them credence. We need to collect evidence to make sure the coincidence isn't spurious. If the test fails, then we should cast the coincidence aside.

This advice seems to make good sense, particularly if we're afraid of making mistakes. Think back to the diagram with the up and down arrows. This advice to test coincidences is what we'd expect from someone with a down-arrow attitude, someone with little tolerance for mistakes.

The advice to carefully test coincidences, however, puts too much faith in our ability to do the testing. Before we commit ourselves to reject any coincidences that don't measure up, let's examine a few cases that illustrate some limits of deliberate testing. We're going to see that even in the face of contrary evidence, we're sometimes justified in placing our faith in the coincidences we observe. The evidence can be wrong.

GIVING YOURSELF AN ULCER

Consider the strange journey of Barry Marshall, the Australian physician who, after discovering in 1982 that a chronic infection causes ulcers, was ostracized by the medical community for the next decade. Before Marshall's findings, the medical community believed that ulcers were caused by stress. Not until 1994 did physicians accept Marshall's claim that

Helicobacter pylori causes ulcers and stomach cancer. As a result of Marshall's research, physicians who used to treat ulcers by cutting out the lower third of the stomach now simply apply antibiotics to clear up the infection. In 2005 Marshall won a Nobel Prize for his findings. In his acceptance speech he quoted the historian Daniel Boorstin: "The greatest obstacle to knowledge is not ignorance; it is the illusion of knowledge."

When a hole forms in a peptic ulcer—an ulcer in the stomach—stomach acid from meals can get into the ulcer and create severe pain. Ulcers are not only painful; they're also life-threatening. Severe cases lead to bleeding. If the ulcer penetrates the stomach wall, the contents of the stomach can leak into the peritoneal space and the patient can die from peritonitis.

Before Marshall's discovery was accepted, physicians typically had two ways to treat ulcers. One technique used surgery to cut off the lower third of the stomach and attach it to the small intestine. The surgery was extreme but effective. Because of the prevalence of ulcers, gastric surgery was a big business at the time. However, 10 percent of patients who had the surgery became gastric cripples who lacked appetite and never completely recovered their health. The second strategy used antacids. The drug companies were making several billion dollars a year from Tagamet and Zantac. In the 1980s, about 3 percent of American adults carried Tagamet with them.

In 1981, Barry Marshall was in the third year of specialized training to gain more experience in cardiology and open-heart surgery. To satisfy a research requirement, Marshall started working with Robin Warren, a pathologist who, two years earlier, had noticed that the gut could be overrun by a corkscrew-shaped bacterium that had to be hardy to survive in the acid-filled stomach environment. The infections coated the stomach. Warren had found them in twenty patients who'd been biopsied because their physicians thought they might have cancer. Warren didn't find cancer, but he did find the bacteria. He wondered if this coincidence meant anything.

For his research project, Barry Marshall agreed to probe this coincidence by investigating these twenty patients to see if they had anything wrong with them. As he looked over the names on the list, Marshall found one of his own former patients; she was a middle-aged woman who had complained to him of nausea and chronic stomach pain. Nothing had showed up from the usual tests, so she was sent to a psychiatrist, who put her on an antidepressant. And here she was on Robin Warren's list, a coincidental link between chronic stomach pain and the corkscrew bacterium.

Another patient on the list was an eighty-year-old man with severe stomach pain. He was diagnosed as having a narrowing in an artery in his gut. He was too old for surgery, so the physicians gave him an antibiotic for the bacterial infection and sent him home. Two weeks later he came back in high spirits and claimed that his stomach pain was gone. Marshall was also struck by that coincidence.

Marshall and Warren began to take these coincidences more seriously. They identified the bacteria as *Helicobacter pylori.* Marshall reviewed the medical archives and found several reports of a spiral gastric bacterium dating back to 1892. No one had yet commented on the coincidence of these reports. Marshall and Warren wondered if *H. pylori* caused some of the stomach problems patients complained about. They submitted their findings to the yearly conference of the Gastroenterological Society of Australia, but the reviewers turned their paper down, rating it in the bottom 10 percent of those they received in 1983.

Marshall wasn't easily discouraged. He prepared a proposal for a proper study, a clinical trial with one hundred patients with duodenal ulcers, to inspect the lab cultures of their stomachs and count how many were also infected with *H. pylori.* If Marshall and Warren were right, many of these ulcer patients, perhaps most of them, possibly all of them, would show *H. pylori* in their stomachs. Such a result would demonstrate a link between the ulcers and the bacterium. The result would open the way for more research, possibly establishing that the bacterium, not stress, causes ulcers.

Marshall and Warren got funding for a year to perform their study, starting in 1982. And it was a bust, at least at first. They didn't find any *H. pylori* at all in the first thirty patients.

So the evidence was clear. Patients with duodenal ulcers don't have *H. pylori.* There should have been no need to waste time and energy completing the study. The coincidence didn't hold up.

Still, they had gotten funding for the study, and Marshall continued. And then, by accident, he found something. One day the lab technicians called him with news that they had gotten a positive culture for *H. pylori.*

Unknown to Marshall, the lab technicians had been throwing out the cultures after only two days. That was the technicians' standard practice in culturing for strep infections, the most common cultures they did. The reason for discarding the samples after two days was that strep cultures are gathered via a throat swab. With a strep infection, on the first day something might be seen but by the second day it would be contaminated by other organisms present in the mouth and not worth saving. However, it turns out that *H. pylori* grows more slowly in a lab culture than strep infections. The cultures taken from the stomach are not contaminated by other organisms, in contrast to the throat swabs. Therefore, the laboratory had been discarding Marshall's samples too quickly.

Fortunately for Marshall, and for ulcer sufferers worldwide, the hospital routine had gotten disrupted. Part of the way through the Marshall and Warren experiment, the hospital detected a superbug and started taking surveillance cultures on every staff member who had been near the infected ward. As a result, the microbiology lab didn't have time to get to the latest culture of Marshall's sample. The lab didn't have time to test and then quickly discard the sample. That culture had been taken on a Thursday, and it sat around for five days, enough time for the *H. pylori* to grow. That was when the lab technicians contacted Marshall with the good news.

Marshall was elated but furious. Six months had been wasted. Marshall and Warren told the technicians to let the cultures grow longer. In the rest of the sample they identified thirteen patients with duodenal ulcers. Every one of them had *H. pylori*.

With hindsight, we can see that the evidence from the first thirty patients was flawed. Those cultures had been discarded too quickly, before they could test positive for *H. pylori.* Examples such as this illustrate why we shouldn't put too much faith in the evidence. We don't know all the conditions that can affect and contaminate evidence, even if the data samples are carefully collected.

Marshall and Warren went on to link *H. pylori* to stomach cancer. Everyone who got stomach cancer also had gastritis. And no person without *H. pylori* had gastritis. That isn't proof, but it's a pretty powerful coincidence.

Marshall and Warren inferred that *H. pylori* generated gastritis, which sometimes led to stomach cancer. They claimed that ulcers (and stomach cancer) are caused by *Helicobacter pylori,* not by stress. If they were right, ulcers should be treated by antibiotics to knock out the infection, rather than by surgery. But the medical community lampooned Marshall. Medical researchers were sure that stress caused ulcers.

In desperation, Barry Marshall ran the ultimate study—on himself. He took some *H. pylori* from the gut of a patient with gastritis, stirred it into a broth, and gulped it down. Within a few days he developed gastritis, moving toward an ulcer. After five days he was waking up, running to the bathroom, and vomiting. After ten days he biopsied his own gut and cultured *H. pylori.* He had personally demonstrated that *H. pylori* causes ulcers. Marshall collected additional data, and by 1983 he had a conclusive case. Unfortunately, the medical community still rejected his papers and his evidence for another ten years.

The tide turned when Marshall moved to the United States and had better access to publicity. His self-experiment was irresistible to magazines such as *Reader's Digest* and the *National Enquirer*. The stories had titles like "Guinea-Pig Doctor Experiments on Self and Cures Ulcer." He lost some scientific credibility but got his message out, and the Food and Drug Administration (FDA) and National Institutes of Health (NIH) took it from there. The popular culture still retains the idea that stress is what causes ulcers, but the medical community has finally stopped doing gastric surgery on ulcer patients.

This case shows why the reasonable conclusion about gathering evidence may not be so reasonable after all. If this example seems like an anomaly, here is another one, a case in which the medical community almost missed the cause of yellow fever because researchers trusted the data too much.

MOSQUITOES

When Walter Reed was sent to Cuba to determine what to do about yellow fever, he knew that the medical community was convinced that unsanitary conditions caused the disease, which spread either by direct contact or, more likely, through the air. Reed's supervisor told him not to bother with the mosquito hypothesis because that had been disproven.

A French-Scottish physician, Juan Carlos Finlay, who had worked in Cuba for decades, had formed the mosquito hypothesis based on a coincidence. Whenever there was a certain type of mosquito—the *Culex* (now called the *Aedes*), there was yellow fever. And the inverse was also true—no *Culex* mosquito, no yellow fever. When the temperature got too cold for the mosquito, yellow fever disappeared, only to reappear the next summer when the temperature rose and the mosquito returned. The *Culex* didn't tolerate higher elevations, and the people who lived at higher elevations escaped yellow fever. So in 1881, Finlay proposed that yellow fever was spread by mosquitoes rather than by unsanitary conditions or physical contact.

Finlay's mosquito hypothesis was widely ridiculed, just as Marshall's *H. pylori* theory of ulcers would be one hundred years later. Finlay, dubbed the Mosquito Man, was called a crank and a crazy old man. Finlay persisted. He did studies letting mosquitoes bite people with yellow fever, then bite healthy volunteers. However, none of the healthy volunteers got sick. Therefore, the medical community, which had been skeptical of the mosquito hypothesis, concluded that the mosquito hypothesis was disproven. The evidence was clear.

After the Spanish-American War ended, Walter Reed, a U.S. Army physician, went to Cuba in 1899 and 1900 to study yellow fever. While in Cuba, Reed heard about a strange case of a jail cell where all the prisoners shared food, accommodations, and bedding, and one of them, only one, came down with yellow fever. Each prisoner contributed in his own way to the common smells; all of them breathed the same air, the same foul smells, yet none of the others got sick. The cell had a small window, and only something like a mosquito could get in.

Also, there was a report from the southern United States about yellow fever outbreaks. A public health service physician in Mississippi, Henry Rose Carter, noted an incubation effect. If a ship with yellow fever victims landed, there might be some cases immediately, but then none occurred for another twelve days to three weeks. Carter wondered about a twelve- to twenty-one-day incubation period for new cases of yellow fever to develop. He speculated that those who were sick when the ship arrived in port were beginning to develop the disease. The others, who got yellow fever two weeks later, might have been bitten by mosquitoes that had first feasted on the initial ship victims and then had time for the disease to mature inside them before they bit fresh victims. The two-week lull was the time it took the disease to develop inside the mosquitoes.

Carter tested his ideas by surveying houses that had harbored a yellow fever victim. After the victim came down with the disease, people who visited the house in the next two weeks remained healthy. Visitors who arrived later risked catching the disease, even if the patient was no longer there. Carter reasoned that if the mosquitoes needed a period of incubation before they became infectious, it would explain why Carlos Finlay's efforts to use mosquitoes to spread yellow fever had ended in failure.

By a fortunate coincidence, Carter was posted to Havana as the harbor quarantine officer. He had a chance to influence Walter Reed with his ideas about an incubation period. Despite the warning from his superior officer, Reed started wondering about Finlay's mosquito hypothesis.

While Walter Reed was on a trip back to the United States, two of his assistants, Jesse Lázear and James Carroll, with Reed's permission, tried an experiment on themselves. They let a mosquito bite someone with yellow fever, waited twelve days, and then let the mosquito bite them. This was the experiment Finlay had tried except that they added an incubation period.

Now the experiment worked. It worked too well. Both Lazear and Carroll came down with yellow fever. Lazear got it so severely he died. James Carroll was delirious (the nurses reported he was saying crazy things such as claiming that he got it from a mosquito) but recovered. Carroll told Walter Reed about the demonstration when Reed returned. They started more controlled experiments and demonstrated that *Culex* mosquitoes indeed transmitted yellow fever.

To make the discovery, Reed's research team had to recover from bad data (the study Finlay did) and from a wrong belief (that mosquitoes are not the culprit). The mosquito theory had to overcome its lack of plausibility; it seemed bizarre that a tiny mosquito could kill a full-grown man. It had to overcome a rival theory: miasma theory asserted that diseases such as yellow fever were caused by unsanitary conditions and unhealthy air. Miasma theory was much more plausible than the mosquito theory. You could smell the stench of a dirty area and imagine that it was unhealthy.

The ulcer example and the yellow fever example show how flawed data can appear to "disprove" an accurate insight. We shouldn't take the ridiculous position of believing in coincidences regardless of the evidence, but we shouldn't automatically believe the evidence either. Evidence can be tainted by variables that we are not aware of.

The ulcer and the yellow fever examples also illustrate the topic of contradictions. Both Barry Marshall and Walter Reed's research team had to cope with contradictions between beliefs and evidence. They had to reject the prevailing wisdom of their time. Of the 12 cases we coded for coincidence, 8 of them also coded for contradictions. That leads to the fourth suspect. Can contradictions lead to insights?

CHAPTER FIVE

Contradictions

CONTRADICTION INSIGHTS SPARK THE EMOTIONAL REACTION "No way!" We give this almost involuntary expression of disbelief when we encounter ideas that just don't make sense. Think of this reaction as akin to a "Tilt!" message when an old-time pinball machine got jostled too aggressively. These mental Tilt! reflexes are the opposite of connections and coincidences. Instead of reacting to the way ideas fit together, we react to inconsistencies.

Contradiction insights send us on the road to a better story. They signal that there's something seriously wrong with the story we're currently telling ourselves. The example of the young cop who spotted a driver ashing his car goes from zero to sixty, from the Tilt! reflex to the new story, in seconds. The Harry Markopolos example took much longer to arrive at the new story that Madoff was running a Ponzi scheme.

Contradictions are different from curiosity insights. Curiosities make us wonder what's going on, whereas contradictions make us doubt—"That can't be right." Only one of the nine curiosity insights also coded for contradictions.

The contradiction form of insight surprised me. I resisted it for a long time. Whereas I knew that many insights spring from connections between ideas, and I also found a bunch of cases that began with a person noticing a coincidence or a curiosity, I wasn't expecting insights to emerge from contradictions. Yet contradictions turned up in more than a third of my sample, 45 out of 120 cases. Even when they overlapped with some of the other themes, such as the connections and the coincidences, they usually were a dominant part of the process. The Tilt! reflex is a powerful trigger.

BANKING ON CONTRADICTIONS

Insights resulting from contradictions paid off for those investment managers whose Tilt! detectors went off from 2003 to 2007 as the bubble in the housing market expanded. While the rest of the financial world merrily went along expanding its investments in subprime mortgages and in the burgeoning housing market, a few investors responded with "I don't believe it!" They couldn't imagine how housing prices could continue to rise at such dramatic rates. They concluded that the United States was gripped by a housing bubble about to burst.

Investment managers weren't the only ones who identified the growing housing bubble. For example, Sheila Bair warned about subprime mortgages soon after she became head of the Federal Deposit Insurance Corporation (FDIC) in 2006. Joshua Rosner, an analyst with a small financial research group, had predicted a vicious cycle of lower home prices and increased foreclosures in a paper he published back in July 2001. Dean Baker, the codirector of a research center in Washington, described the disastrous effects of the eventual collapse of the housing bubble in an article he published in 2002.

I selected the following five cases to illustrate different cues that set off the Tilt! reflex as the bubble was expanding. Each route started when a person spotted a contradiction.

Steve Eisman was an insider. After getting a law degree and discovering that he hated being a lawyer, he went to work for Oppenheimer Securities, a small Wall Street firm. In 1991, he was appointed the lead analyst at Oppenheimer for the first subprime mortgage lender to go public. Eisman had been specializing in subprime mortgage companies for more than a decade by the time of the 2007–2008 crash. He was well-known for being arrogant and opinionated—and often right. Even his wife, who is devoted to him, admits that he is beyond outspoken—not tactically rude in order to gain an effect but genuinely rude.

Eisman at first was an advocate for the subprime mortgage market, but he changed his mind when he saw the shoddy loan practices in the late 1990s. He didn't make any friends when he called attention to these practices, but he was vindicated when the subprime mortgage market experienced an earlier collapse in the late 1990s.

That collapse sent a number of subprime lenders into bankruptcy. Eisman expected that they had learned their lesson and would become more careful in the future about giving mortgages to people with marginal ability to keep up their payments, but that isn't what happened. By 2005, the subprime mortgage industry returned bigger than ever. Eisman again watched delinquency rates go up and the quality of the loans go down. Financial institutions were again lending money to homeowners who were less able to demonstrate their capacity to repay even though the previous delinquencies should have made the lenders tighten their standards. In 2005, Eisman saw the subprime growth rates continue to rise even as the interest rates were rising, a fact that should have cooled off the demand.

When Eisman investigated, he found that the lenders had learned a lesson from the 1998 subprime crash. It was just a different lesson than he expected. When the subprime bubble burst in the late 1990s, the lenders went bankrupt because they had made bad loans to unworthy applicants and kept too much of those loans on their books. Instead of learning not to make bad loans, the industry had learned not to keep the risky loans on their books. They sold the loans to Wall Street banks that repackaged them as bonds, found ways to mask the riskiness, and sold them to unwary investors. That's when Eisman's Tilt! reflex went into overdrive. It reached even higher levels when he investigated the ratings agencies. In a conversation with a representative of Standard and Poor (S&P), Eisman asked what a drop in real estate prices would do to default rates. The S&P representative had no answer because the model the agency used for the housing market didn't have any way to accept a negative number. Analysts just assumed housing prices would always increase.

By this time, Eisman had started his own hedge fund. He was certain that the subprime market would blow up again, and he was determined to find a way to short the market and profit from Wall Street's blind greed.

John Paulson, owner of a struggling hedge fund, bet against the subprime market and collected more than $15 billion, the world's biggest payday. Unlike Eisman, Paulson and his partner, Paolo Pellegrini, were outsiders to the housing and subprime markets. Their Tilt! detectors went off as they watched the trends in those markets. When the price of a commodity increases beyond its worth, smart investors sell their holdings to lock in their profits. But when the prices increase for no apparent reason, and the rate of increase itself increases, there might just be a contradiction of supply and demand. There might be a bubble because the demand itself creates more demand as investors expect the upward trend to continue. In 2005, when Paulson and Pellegrini ran the numbers about the growth rate in the subprime market, they concluded that the increases in real estate prices weren't sustainable. Yet the whole enterprise depended on continued price increases. The prices didn't have to fall. The bubble would quickly collapse even if the prices leveled off. For Paulson and Pellegrini, the contradiction was between the marketplace belief that real estate prices would continue their dramatic rise and the economic forces that prevent housing prices from rising forever. It was a question of *when* the bubble would burst, not *whether* it would burst.

Michael Burry was also an outsider, a defiant outsider who was skeptical about the major investment firms. Burry spotted the bubble earlier than most, perhaps because of his nature as a contrarian. He was socially isolated as a boy, eventually diagnosed with Asperger's syndrome. Instead of socializing, he spent his time studying, and upon graduation he was admitted to medical school. In 1997 when he got his medical degree, he'd accumulated a tuition debt of $150,000.

While doing a residency at Stanford, he put a personal ad on an online dating service: "I'm single and have one eye and a lot of debt." (His left eye had been removed after he was diagnosed with retinoblastoma—a rare form of cancer—when he was one year old.) This bizarre come-on attracted a young corporate finance specialist who appreciated Burry's honesty, and they were married three weeks later.

Burry had solved one of his problems, a longing for social connection. Another major problem was his medical school debt. A third problem was that he didn't enjoy medicine very much. Multiple problems can sometimes combine into a solution, and that's how it turned out for Burry. For many years he'd pursued a hobby of following the stock market, and when he was twenty-nine, he decided that he was more interested in financial investment than in medicine. If nothing else, it would let him pay off those loans. Burry's family provided him with some capital to get started. One account has Burry's wife raiding her retirement account to provide him with capital. Presto, Burry had his own hedge fund.

Burry's website on financial trends had attracted a small following, and some of his fans were happy to buy shares in his new company. Because of his analytical skills, by 2003 Burry's hedge fund was managing $250 million and he was making $5 million a year. And that's when he got interested in subprime mortgages.

His suspicions were aroused in 2003 as the subprime market started its rise. When he dove into the data in 2004, he found that lending standards had been declining. But why should standards have to decline if people really believed the housing market would always keep growing? One top-rated analyst intoned, "Home prices have never gone negative." Another proclaimed, "Home prices won't even go flat."

If they believed their hype, they'd be tightening their standards, not loosening them. That inconsistency bothered Burry. If housing prices could be counted on to always go up, why did the lending institutions have to go after less qualified applicants? Why did these institutions have to rely so heavily on variable-interest-rate loans that lured homeowners with low initial rates that were not fixed and were bound to increase? Burry sensed a contradiction between the public display of optimism and the private practice of reducing standards. He predicted that the real estate market would soon decline because many of the people taking out the subprime mortgages wouldn't be able to keep up with the payments. Burry estimated that the housing market had a two-year horizon. All the risky loans made at teaser rates in 2005 would become much riskier once the two-year window closed in 2007 and those low, teaser interest rates increased to 11 percent. Burry anticipated that starting in 2007 the default rates would sharply increase, and the bubble would burst. Who were the homeowners taking out these loans and accepting a risky option to pay nothing down? People who didn't have much savings. Not the kind of people who were well suited to handle interest rates of 11 percent.

Greg Lippmann was an insider. He bought and sold subprime mortgage bonds at Deutsche Bank and stumbled upon the financial bubble in 2005 when his bosses directed him to bet against the housing market. Lippmann didn't like the idea—why fight a market that was always going to rise? He dutifully studied the housing situation, and what he uncovered horrified him. Even though housing prices were still rising, the mortgage default rates had increased from 1 percent to almost 4 percent. He calculated that if the default rates rose to just 7 percent, the lowest-rated mortgage bonds would become worthless. If the default rates reached 8 percent, the next level of bonds would become worthless, and so forth. Tilt! The increased default rates were bound to increase supply and reduce demand. They were bound to cool off the housing market. These implications frightened Lippmann. If housing prices just leveled off (let alone declined), it would devastate financial markets that depended on a continual increase.

Gene Park was another insider. He worked for the insurance company AIG (American International Group). In 2005, he read a *Wall Street Journal* article on New Century, a mortgage lender with a high dividend, and wondered if he should invest some of his own money in New Century. He found, though, that the company's success in the mortgage business stemmed from subprime loans that seemed frighteningly poor to him. Park also saw an increase in risky investments involving insurance on bundles of mortgages. The insurance was designed to reduce risk by spreading over lots of mortgages but didn't give much weight to the risk that the entire U.S. housing market might fall. Then Park got a call from a penniless and jobless friend who mentioned that he was getting offers from banks, each wanting him to buy a house he couldn't afford. That's when Park's Tilt! detector went off. He realized that there was an unsustainable investment bubble around subprime mortgages. And his company, AIG, was on the wrong side of the bubble.

Unlike the other investors, Park didn't have any way to turn his Tilt! response into a windfall. He couldn't even convince his bosses that AIG was in big trouble. The other four reaped enormous profits for themselves and for their companies. For example, Burry made $750 million in 2007 for the investors in his hedge fund. Paulson did even better. Their gains depended on additional insights about how to use special investment tools to buy insurance on companies they expected to fail.

For the purposes of this chapter, however, I am just interested in the way they arrived at the conclusion that Wall Street, the financial engine of the United States, was headed for a dramatic and imminent breakdown. Most financial experts seemed oblivious to this breakdown. It also took business and government leaders by surprise. We seem to be better prepared to spot asteroids that threaten to collide with our planet than to foresee this type of financial devastation. Asteroids are a known threat that is amenable to scientific tracking. Disasters like the 2007–2008 financial breakdown are less visible because they are conceptual rather than physical, and because they violate so many strongly held beliefs. To quote Mark Twain, "You can't depend on your eyes when your imagination is out of focus."

Five different investors, some insiders, others outsiders, drew on different perspectives. They used different data to pick up the contradiction between the euphoric expansion of the subprime mortgage market and their own realization that this market was about to collapse.

All five were skeptics. We are told that to make discoveries, we have to keep an open mind. However, a suspicious mind can also pay off. These five investors, and Markopolos, the Bernie Madoff sleuth, tended to be suspicious, if not cynical. Their skeptical mind-sets helped them investigate pathways that others missed.

Their skepticism spurred me to become curious about a skeptical mindset as opposed to an open-minded one. I went back and recoded the data, looking at the 45 cases of contradiction insights in my sample. In two-thirds of the contradiction instances, the person gaining the insight had a suspicious mind-set rather than an open mind. He or she resisted the popular enthusiasms.

Meredith Whitney illustrated this skeptical stance in the way she tracked down the problems facing the Wall Street firm Bear Stearns. Whitney was a managing director at Oppenheimer until she left in February 2009 to start her own advisory firm. She'd been trained at Oppenheimer by Steve Eisman in the mid-1990s.

Whitney had flagged the growing subprime mortgage problem. Her research report in 2005 warned about scandalously reduced credit standards; subprime lenders were facing huge losses that threatened a round of corporate insolvencies. Unfortunately, few people paid attention to Whitney's warning.

The financial community did listen to Whitney on October 31, 2007. That's when she made her reputation. She publicly predicted that Citigroup was facing a financial crisis and would have to cut its dividend or face bankruptcy. At the time, Citigroup seemed to be in good shape. Once Whitney exposed its weakness, Citigroup shares dropped 8 percent by the end of that afternoon. Its CEO resigned four days later. And it cut its dividend two weeks after that.

In early 2008, Whitney decided to take a closer look at Bear Stearns. Two Bear Stearns hedge funds had abruptly closed down in mid-June 2007, but the company seemed to weather these losses and listed a strong asset sheet. The failed hedge funds had invested heavily in the subprime mortgage market; most analysts assumed that Bear Stearns had the resources to ride this crisis out.

On March 8, 2008, after Whitney heard rumors about the solvency of Bear Stearns and Lehman Brothers, she learned that the Federal Reserve was going to make it easier for investment banks to borrow money. "My first reaction was, 'Oh, this is going to be completely fine.'"

But then she heard that the investment banks that needed these funds would not have access to the loans for several weeks. She called back the traders who had previously complained about Bear Stearns, but now they clammed up and acted as though there were no problem at all. They insisted that they were still trading Bear Stearns stock.

Whitney immediately felt they were lying to her. "When people that you deal with all the time outright lie to you, then you know there's serious panic there." She contacted another of her sources, asking directly about Bear Stearns's solvency, and she was told it was possible Bear Stearns might go under. Unfathomable, but possible.

Whitney has a reputation as one of Wall Street's most skeptical research analysts, and she began to look through that lens at what was happening to Bear Stearns. She deliberately adopted a suspicious mind-set and tried to interpret every data point as evidence that Bear Stearns was finished. In short order, she convinced herself. She knew that many of Bear Stearns's businesses were contracting. The other investment banks were taking market share from Bear Stearns's prime brokerage business as hedge funds withdrew their cash. She knew that Bear Stearns was highly leveraged, at least 30 to 1, sometimes 50 to 1. She thought, "This is it—Bear's out. . . . If people start pulling assets, they [Bear Stearns] can't cover their debt service. . . . And then it's over. . . . By Thursday I knew it clear as day. . . . This has happened to me several times . . . where I know things in my gut [e.g., the risks created by the erosion of subprime lending standards, the impending fall of Citigroup]. But I can't believe I'm the only person who is putting it together."

I like the notion that Whitney was deliberately using a skeptical lens. This lens gave Whitney a different perspective than if she had tried to keep an open mind. Her suspicious lens enabled her to gain the insight that Bear Stearns was going to collapse.

Whitney acted as a detective, going against popular beliefs to gain insights that eluded her more open-minded colleagues. The same was true of the five contrarian financial investors who saw the end of the subprime bubble. An open mind has its virtues, but a suspicious mind provides its own unique benefits. It produces a different view and exposes alternative facts and trends.

A skeptical stance helped John Snow in this next incident. Once he lost his faith in miasma theory, he knew where to look to find the contrary evidence and to lend support to his theory that cholera was being spread by contaminated water.

THE BROAD STREET PUMP

The Tilt! reflex led to John Snow's discovery about cholera in the mid-1800s. Cholera strikes in epidemics as the infection spreads from one victim to another. The cholera microbes cause diarrhea and vomiting that rapidly dehydrate their victims, quickly killing them. People can die less than a day after showing the first symptoms. Cholera made its first appearance in Britain in 1831. By 1833, when the epidemic ended, more than 20,000 people had perished in England and Wales. The next epidemic of 1848–1849 killed another 50,000.

At the time of Snow's discovery most people believed in miasma theory, that diseases such as cholera and yellow fever were spread by bad air and disgusting smells created by unsanitary conditions. Those were the days before anyone had a germ theory of disease. There were no instruments for detecting microbes. Diseases such as cholera spread terror because they were so lethal and so mysterious. People didn't know how to protect themselves.

John Snow was born in 1813 in York, England, to a working-class family. He became a surgeon's apprentice and then moved to London, where he eventually received a degree as a medical doctor. He published articles on a range of medical and public health topics.

Snow became prominent in London for his mastery of anesthesia. Medical and dental procedures had been brutal affairs. A good surgeon needed a sharp saw and muscular assistants to hold the patient down. Then in October 1846 a Boston dentist reported that he used a gas—ether—to reduce pain by making the person unconscious during the procedure. A London dentist adopted ether as a remedy for pain shortly thereafter, and Snow observed a demonstration at the end of December.

One problem with anesthesia is getting the dosage right. Too much and the life of the patient is endangered. Too little and the patient might wake up in the middle of the procedure.

Snow set out to understand how to safely regulate the dosage. Dentists had administered ether by soaking rags in it and then clamping the rags over the patient's nose and mouth. Snow designed an inhaler that controlled the amount delivered more precisely than the rag method. Using studies with small animals, and then with his own patients, he determined how the amount of ether affected the length of time the patient remained unconscious. He even used himself as a subject. He'd write down the time, give himself a dose of anesthetic, pass out, wake up a few minutes later, note the time he regained consciousness, and calculate the effect of the dosage. When chloroform replaced ether as an anesthetic, Snow shifted his research and made himself an authority on chloroform as well.

His research and his monographs helped pioneer the use of anesthesia to reduce pain in medical procedures. He was brought in to administer chloroform to Queen Victoria to help her with the birth of her last two children. She reported that she was gratified by the effect of the anesthetic.

So when Snow gained his insight into how cholera was spread, he wasn't someone who could be treated as a crank, the way the SEC regarded Harry Markopolos or the Australian medical community treated Barry Marshall. Snow was an eminent physician. Yet his reputation didn't spare him from widespread criticism. When people contradict the prevailing wisdom, even professional prominence won't protect them.

Snow became interested in cholera after reading about a contradiction. A sailor died of cholera in a lodging house. A few days later another person checked in to the same room and contracted cholera. That sequence didn't fit with miasma theory. If inhaling noxious air caused cholera, why didn't other people in the lodging house or the neighborhood get sick? Noxious vapors should spread in currents, afflicting everyone in their path, but cholera didn't follow such paths. The sailor story activated Snow's Tilt! reflex.

Now he wondered about other pieces of the puzzle: people who were in the same room with victims sometimes came down with the disease even if they didn't touch the victim, and others who touched the victim sometimes didn't get it. That pattern of observations also seemed to contradict miasma theory.

Another contradiction struck Snow. As a physician, he expected cholera victims would show lung damage if cholera were transmitted by bad air. Yet the victims' lungs seemed normal. The damage was in their digestive systems, a coincidence that implied a connection—that they caught the disease from something they ate or drank.

That was Snow's insight. Cholera was caused by something the victims ingested, not by what they breathed. But what had they ingested? He gathered more data and began to suspect that cholera was spread through ingesting the waste matter of other victims, typically by drinking contaminated water.

Cholera seemed to parallel water systems. Snow found one outbreak in 1849 in which twelve people died; they all lived in one slum, all in one row of connected cottages, and they all drank from the same well. Plus there was evidence that sewage water could get into that well. Another set of nearby similar flats used a different well, but only one person died of cholera in the same epidemic. Surely they breathed the same air. The evidence pointed to the water system.

Snow also found a natural experiment: a community that got water from two different companies, an upriver company and a downriver one. The upriver company drew its water from the Thames north of London and, more importantly, upriver from a sewage drainage point. The customers who used this company didn't get cholera. The downriver company drew its water below the point where raw sewage entered into the Thames. Those who used this water company showed high rates of cholera. All the citizens of the community breathed the same air. So there was a nice coincidence. Snow relied on contradictions and coincidences as he built his case that cholera was caused by contaminated water.

In 1854 the United Kingdom was hit with another big cholera outbreak. John Snow and an allied investigator, the Reverend Henry Whitehead, showed that cholera cases clustered around one specific water pump, the Broad Street pump, in Whitehead's district. The following figure shows how the cases clustered around the pump, which is marked by an X. Snow speculated that the water source for this pump had become contaminated.

Snow noted other anomalies during this outbreak: none of the workers at a local brewery came down with cholera, yet they breathed the same air as everyone else. But unlike everyone else, the brewery workers slaked their thirst with beer, not water.

Snow tracked down an important piece of data: a woman who had moved away from the area of the Broad Street pump still got cholera.

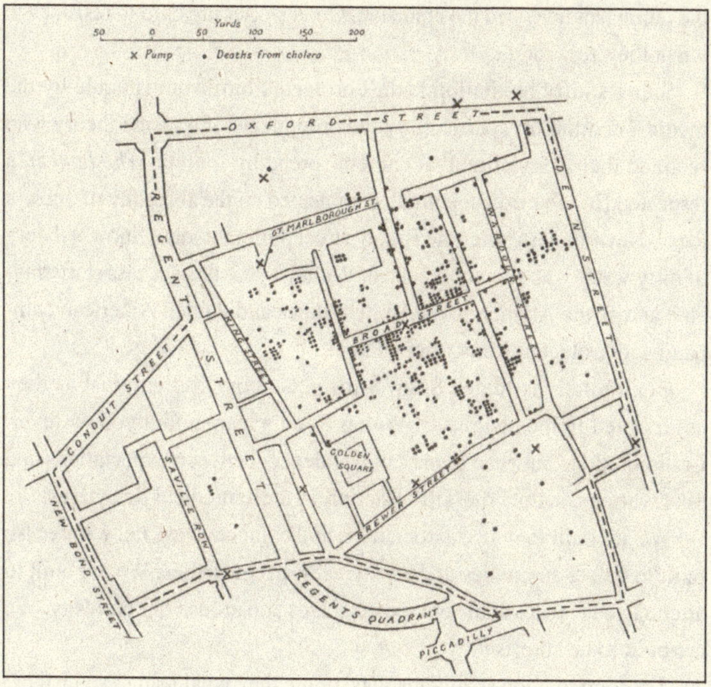

Her case should have disproven Snow's ideas, except that shortly before her death her son had filled up a jug from the Broad Street pump and brought it to her because she liked the taste of that water so much. Instead of challenging Snow's idea, this piece of evidence supported it.

Later investigation revealed that the Broad Street pump had gotten contaminated through the dirty diapers of an infant. The infant's mother cleaned the diapers in a bucket and tossed the fouled water into a cesspool that ran into some cracks in the pipes of the Broad Street pump. The local council decided to seal up the Broad Street pump by removing its handle. People from the neighborhood, who didn't believe Snow's dirty water theory, were angry because the water from this well tasted so good, and then the cholera epidemic ended. The epidemic was already waning, so removing the pump handle didn't make much difference. Nevertheless, the public seized on the coincidence. The popular story is that removing the pump handle ended the epidemic. Good stories are hard to resist, even when they're wrong.

Snow's lofty reputation hadn't protected him from ridicule by the medical community, particularly the proponents of miasma theory who retained their conviction that cholera spread by foul air. *The Lancet,* a respected British medical journal, commented on the absurdity of Snow's ideas. Nevertheless, after the Broad Street pump incident Snow's theory of dirty water was quickly accepted. Within a year this success story traveled across the Atlantic to the United States and guided American campaigns to protect citizens from cholera.

Contradictions fueled Snow's insights about cholera, just as they contributed to the insights of Walter Reed's team about yellow fever. Contradictions between theory and evidence, between expectations and observations, set off the Tilt! reaction in these medical detectives.

We are built to notice associations and coincidences, and we are also built to detect anomalies, inconsistencies, irregularities. We are built to attend to cues that violate our expectancies and arouse our curiosity. We are built to be surprised.

Are contradictions really insights, or are they what lead to an insight? They certainly lead to insights, to the better story. I believe they also count as an insight. Noticing a contradiction shifts the way we understand things. Markopolos knew something after scanning Madoff's results that he didn't know before, even though he hadn't yet sorted out what Madoff was doing. Snow knew that miasma theory didn't explain cholera, even before he found the real cause.

Thomas Kuhn, a physicist who theorized about the history and philosophy of science, wrote *The Structure of Scientific Revolutions* to describe how anomalies lead to breakthroughs. Kuhn wanted to explain the origins of breakthroughs such as Copernicus discovering that the sun, not Earth, is the center of our solar system. Or Roentgen discovering X-rays, a phenomenon others had observed but dismissed because the notion of X-rays was too inconsistent with their beliefs. Kuhn argued that most research is "normal science"—studies that add more detail to existing knowledge and theories. Normal science, however, usually suppresses the contradictions, the observations that don't fit the frameworks that the scientific community shares. Over time these discrepancies grow into crises until someone comes along to propose a paradigm shift, a new way to understand natural forces, a new set of questions, a new way to search and research. Kuhn described these paradigm shifts as scientific revolutions. They require paradigm destruction—the shedding of the previous paradigm. The new paradigm changes how scientists understand phenomena. It changes what scientists see and how they act in designing experiments. Paradigm shifts count as insights because the result is a shift from a mediocre frame to one that provides a better understanding of the same phenomenon.

Kuhn seems to me to overemphasize revolutionary paradigm shifts. He is too enthusiastic about paradigm destruction, the discarding of the previous paradigm. Connections, coincidences, and curiosities also can lead to important and revolutionary scientific insights without paradigm destruction. Nevertheless, Kuhn and other philosophers and historians of science have helped us appreciate how new ideas often emerge from a breakdown of previous beliefs. New paradigms emerge from the recognition of anomalies and contradictions.

SEEING THE LIGHT

Albert Einstein's theory of special relativity seems to fit this pattern of contradictions that lead to insights. Kuhn cited Einstein's theory as one of his examples of a paradigm shift.

At the age of sixteen, Einstein began to conduct thought experiments about beams of light. These thought experiments were mental exercises that helped Einstein appreciate properties of light and also helped him notice anomalies and inconsistencies. Einstein imagined different conditions and possibilities, pursuing these speculations for ten years. His thought experiments resulted in his theory of special relativity in 1905 when he was twenty-six.

In one of his thought experiments, Einstein wondered what would happen if he traveled at the speed of light alongside a light beam. What would he see? Logically, the light beam should appear frozen, traveling at the same speed as he, like two automobiles in neighboring lanes driving at the speed limit. But not long before, in 1899, James Clerk Maxwell had published a set of equations to model electric and magnetic fields. These equations predicted that electromagnetic waves would propagate through a vacuum at the speed of light. Therefore, if Einstein rode a beam of light, then a second light beam traveling alongside him would still travel away from him at the speed of light. If he turned on a flashlight, the light would travel away from him at the speed of light. So there was a paradox, a contradiction, because nothing can go faster than light.

Eventually, Einstein resolved the paradox by changing the constant. Previously space and time were thought to be constant. This was the Newtonian view. A second on Earth is the same duration as a second anywhere else in the universe. According to Newtonian physics, we could synchronize our watches anywhere in the universe because time beats uniformly. Likewise, a foot on Earth is the same as a foot in every other location. Space and time should look the same to all observers, regardless of their relative velocities.

Einstein's thought experiments started him thinking that *the speed of light* should be the constant. In exploring the implications of this new belief, Einstein speculated that space-time was variable. Using mental images of clocks and trains, light beams and speeding bicycles, Einstein showed that for the speed of light to be constant, time must appear to pass more slowly the faster you move. Lengths must contract and masses must appear to increase as you approach the speed of light. If you're in a car, moving at 90 percent light speed, everything looks normal. Time seems to be passing normally; everything in the car seems to be the usual size. It's only to others watching from outside the car that time in the car seems to be moving too slowly and lengths appear too short.

Light isn't like a traveler on a moving sidewalk in an airport. If the sidewalk is moving at 2 miles per hour (mph) and you are rushing to catch your plane at 4 mph, you'll be crossing through the terminal at 6 mph. But light doesn't have that property. It always appears to travel at the same speed whether or not it originates from a moving source.

Instead of explaining away the inconvenient contradiction, that even if he were moving at the speed of light, a light beam would travel away from him at the speed of light, Einstein took the contradiction seriously and discovered a way to fit everything else around it.

Einstein, like John Snow, like Harry Markopolos, and like the contrarian financiers, used a contradiction to make a discovery. This strategy of using contradictions seems very different from the strategies of noticing coincidences and curiosities and making connections.

As I waded through the examples and through the data, I enjoyed my investigation because each example of insight was so inspiring, but I grew less confident that I'd be able to formulate an answer to the mystery of what gives rise to the flash of illumination. I had too many answers, and they didn't seem to relate to each other. And there was one more strategy to think about, the strategy most commonly invoked by the scientific community.

CHAPTER SIX

Creative Desperation

Trapped by Assumptions

SOME INSIGHTS ARE ACCIDENTAL, the result of unplanned happenstance or of being in the right place at the right time. Think about the young cop or Martin Chalfie sitting in his lunch seminar. Others are deliberate, arising when people are stuck and need some sort of breakthrough. In his work on chess grand masters the Dutch psychologist Adriaan de Groot used the term "creative desperation" to describe some of the brilliant strategies that players invented when they got into trouble. The clock would be winding down, they'd be in a tough situation, none of the plausible moves worked, and then, out of desperation, they would find an unorthodox line of play that would save them, gambling on some leverage they might never have tried to use if any of the acceptable moves had looked promising.

Almost all the scientific research on insights, as noted earlier, takes place in laboratories using puzzles designed to stymie the subjects. Some of the subjects get stuck and have to give up. Others struggle, feel lost, and then have unexpected insights about the solutions. To research psychologists who have carefully studied the nature of insight, this theme of creative desperation is the epitome of insightful problem solving. It is the paradigm they use in their experiments because there is a solution, a right answer, that is discovered through a flash of illumination rather than steady analysis. These are the kind of situations that Graham Wallas had in mind when he formulated his four-stage theory.

Creative desperation, a fifth insight strategy, is very different from connecting ideas or noticing coincidences or curiosities or contradictions. Creative desperation requires finding a way out of a trap that seems inescapable.

FIGHTING FIRE WITH FIRE

Creative desperation does not, of course, just happen in the laboratory or over the chessboard. Sometimes it is a matter of life and death, as was the case in Mann Gulch, Montana (see photograph), on August 5, 1949, when Wagner Dodge found a way to escape a forest fire. A team of fifteen smokejumpers parachuted into western Montana that afternoon to control a forest fire. Less than two hours later, twelve of the smokejumpers lay dead or dying after being trapped by a blowup—a sudden and unexpected firestorm resulting from a collision of fire and winds. A blowup to a typical forest fire is like a hurricane to an ocean storm.

The team had been dropped at the top of Mann Gulch on an extremely hot day in the midst of a dry season. Their plan was to hike down on the north slope of Mann Gulch (on the left in the photo) to the Missouri River and cross over to the south slope (on the right) where the fire was burning. They wanted to attack the fire from below because fires spread upward and they didn't want to put themselves at unnecessary risk. The Missouri River runs horizontally at the bottom of this photograph of Mann Gulch.

Wagner Dodge, the thirty-three-year-old team leader, was described as taciturn, even by his wife, who greatly loved and respected him. "He said to me when we were married, 'You do your job and I'll do mine, and we'll get along just fine.'" Dodge had been with the Forest Service

for nine seasons before the Mann Gulch fire, and he had been a smoke-jumper foreman for four years. As the leader of this crew, he was supposed to go through a three-week training course with them, but at the last minute he was reassigned to a maintenance task. He and his crew barely got to know each other before being dropped into Mann Gulch.

They hiked most of the way down Mann Gulch, but then Wagner Dodge saw a bit of smoke down at the bottom of his part of the valley, the north slope. He immediately recognized that somehow the fire had "spotted," that is, the fire on the south slope had sent sparks and embers that were carried by the wind to start a new fire, a spot fire, on the north slope, directly downhill from them. A cascade of spot fires can quickly lead to a blowup, and that was about to happen.

Dodge could see the beginnings of a fire way down below, near the river. Dodge knew that this fire was going to come roaring up the north slope of the valley where he and his team were located. It would likely kill them all.

Dodge ordered his team to run for safety back to the top of the hill. However, when he looked back, he realized they wouldn't make it. The fire was gaining on them and was picking up speed. In his book about Mann Gulch, *Young Men and Fire,* Norman Maclean used one of the most heartbreaking graphs I have ever seen to illustrate the plight of the firefighters. My legs get weak just studying the curves.

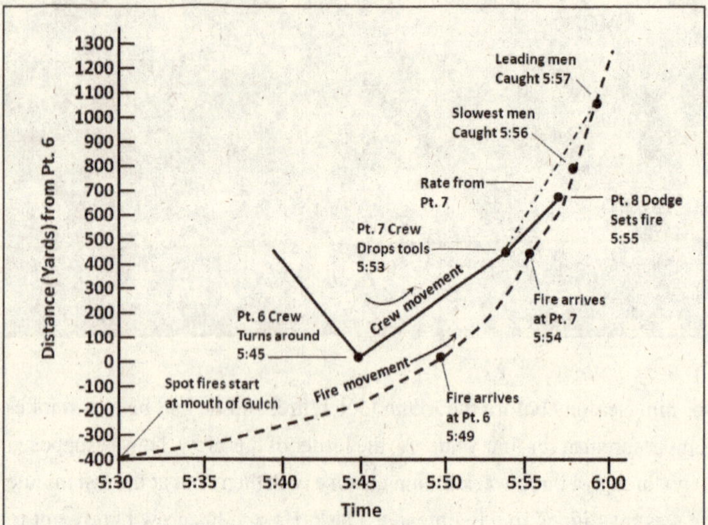

The bottom axis shows the time sequence. Just half an hour passes once the fire crosses over to the bottom of Mann Gulch until it catches the fastest runner. The fire is the dotted line; the smokejumpers are the solid line. The graph shows the race between them. The smokejumpers were descending the valley until they reached Point 6, which is where Dodge saw the first signs of smoke and they reversed their course at 5:45. The vertical axis shows the distance above and below Point 6.

At Point 7, at 5:53, the crew members drop their tools in their mad scramble. The fire is now only a minute behind them, driven by 30 mph winds. The fire is 30 feet high by some estimates, 200 feet high by others. Dodge described it as a wall of flame. It was like a tsunami, made of flames instead of water.

The fire moves slowly at first, not reaching Point 6 until 5:49, nineteen minutes after it spotted at the bottom of Mann Gulch. Unfortunately, the fire picks up speed. As the valley gets steeper, the fire can move faster but the legs of the smokejumpers have to work harder. At its fastest the fire is traveling about 660 feet a minute. The last portion of the graph, on the right-hand side, the end of the race for the smokejumpers, is a 76 percent slope.

Norman Maclean had once been caught in a forest fire himself. He survived a close call and offered his reflections on his experience: "Burning to death on a mountainside is dying at least three times, not two times as has been said before—first, considerably ahead of the fire, you reach the verge of death in your boots and your legs; next, as you fail, you sink back in the region of strange gases and red and blue darts where there is no oxygen and here you die in your lungs; then you sink in prayer into the main fire that consumes, and if you are a Catholic about all that remains of you is your cross."

I had a chance to visit Mann Gulch several decades after the fire of August 1949. I was part of a team investigating the 1994 fatal South Canyon fire in Colorado. Our team met in Missoula, Montana. As a side trip, some of the researchers went to Mann Gulch, taking boats up the Missouri and hiking up the valley. When we got high enough and the canyon got steep enough, I tried running up the 76 percent slope to see what it felt like. At the time I was in moderately good shape—I had been running five miles a few times a week, an eight-minute/mile pace, nothing special. I wasn't even close to the physical condition the smokejumpers were in. As I ran up Mann Gulch, it didn't take me long to become exhausted. I imagined a forest fire at my heels. I imagined the panic, the adrenaline. I still couldn't make it very far.

The slowest smokejumper was caught at 5:56 in Maclean's reconstruction of the event, eleven minutes after the crew reversed its course at Point 6. The others were caught in the next minute. Two smokejumpers did make it to safety, but the other twelve died. Look at the graph again, at the cruel upward slope of the fire.

Wagner Dodge survived through creative desperation—an ingenious counterintuitive tactic. To escape the fire, he started a fire. At 5:55 in Maclean's timeline, Dodge lit a fire in front of him, knowing that this escape fire would race uphill and he could take refuge in its ashes. He wet his handkerchief from his canteen and put it over his mouth and nose, then dived facedown into the ashes of the escape fire to isolate himself from any flammable vegetation. He was saved with less than a minute to spare. But he could not persuade anyone to join him in the ashes. None of the others could make sense of what he was doing. He had invented a new tactic, but he never had a chance to describe it to his crew. As one of the two survivors put it, upon seeing Dodge light a fire, "we thought he must have gone nuts."

I have read several accounts of this incident, but none of them explains Dodge's insight. He invented the idea of an escape fire as an act of creative desperation. So I am forced to imagine where the insight came from.

I imagine Dodge and the crew running for their lives, dropping their tools, realizing they are trapped. I try to imagine Dodge's beliefs about what is going on. Four beliefs in particular seem to anchor Dodge's understanding. One anchor is the uphill slope. It favors the fire, not the runners, and it is getting steeper. A second anchor is the fire behind me. It is picking up speed. Perhaps a third anchor is any island of safety I might find. If I could get to the top of the ridge, the fire would stop racing after me. Or if I could find a rocky patch, the fire might leave me alone because it wouldn't have any fuel. I see a rocky area way up ahead (the other two survivors reached these rocks and took protection there), but I know I can't get there in time. It is about 200 yards away, uphill yards, with the fire perhaps thirty seconds behind me. A fourth anchor is the heavy dry grass I am running through, bunchgrass and cheatgrass. It is the fuel that the fire is eating.

I don't spend any time thinking about how to change the slope of the valley walls or beat the fire back. These anchors are part of my situation, but beyond my control. I can't reach the rocky areas on the ridge before the fire will catch me. But maybe I can do something to this last anchor, to the fuel. How can I neutralize the fuel? Of course! I can burn it! Fire, my enemy, can also be my friend. I light the escape fire to race up the slope ahead of me, and I dive into the hot ashes, my sanctuary from the blowup.

In my imaginary account of Dodge's thinking, he is looking for any assumption that he can overturn. When he finds one, the grass fuel, he has an escape plan.

DISARMING A FLAWED ASSUMPTION

Dodge later testified about his actions but didn't provide an explanation for his insight about the escape fire. So we can never know what went through his mind, but there is an analogous case that seems to follow the same script. In May 2003, Aron Ralston, an American mountain climber, was hiking through some canyons in Utah. He fell into a crevice and his right arm became pinned by a boulder. Ralston was trapped for over five days, and he recounted the ordeal in his book, *Between a Rock and a Hard Place.* James Franco played the role of Aron Ralston in the movie version of the ordeal, *127 Hours*. Franco did a wonderful job of conveying Ralston's frustration and his anger at himself for failing to let anyone know his plans for the trip.

At first, Ralston tried to use a small pocketknife to carve away the rock and free his arm, but with no success. All he accomplished was dulling the blade. After a few days he accidentally cut into the thumb of his trapped arm and confirmed that the trapped part had died from lack of circulation. It was no longer worth saving. It might even be poisoning him. He was out of food, almost out of water, and completely isolated. He fully expected to die.

He gave up trying to free his arm and instead tried to amputate it in order to free himself, but by this time his knife was too dull. It would never be able to cut through the bones of his forearm. He truly was trapped.

At one point he lost his composure and gave in to an emotional outburst. He tried to yank his arm free of the rock that encased it, thrashing back and forth. During this tantrum he felt his right arm bend unnaturally, and that was the hint he needed to make a discovery. He didn't have to cut through the bones of his arm. The boulder had pinned his arm so tightly that he could *snap* the bones. Once he did, he used a smaller blade to cut through the tissue (and, most painfully, the nerve bundle). He managed to free himself and find help.

As soon as Ralston shifted his goal from freeing his healthy arm to freeing himself from his dead arm, he discovered that the boulder was no longer his enemy. It was now his friend. The boulder enabled him to escape by providing the solid leverage he needed to break the bones in his arm. Similarly, by shifting his focus from the fire behind him to the fuel in front of him, Wagner Dodge made fire his friend. I see both cases as deliberate insights achieved by desperate men trying to find any leverage points they could use to save their lives. They both jettisoned the assumptions trapping them.

CHERYL'S KISSES

Deliberate insights rarely have the drama or the terror facing Wagner Dodge and Aron Ralston. More often the desperation resembles a chess player mired in a bad position, rather than a person struggling to avoid imminent death. Many years ago, Cheryl Cain, the financial manager of the research company I owned at the time, needed the employees to fill out time cards at the end of the week. Unfortunately, many people weren't complying. (Confession: I was one of the worst offenders.) Cheryl had to chase us down. Because so much of our work received government funding, she lived in fear of a surprise government visit and audit. The officers of my company issued stern directives to get people to fill out the time sheets, but to no avail.

One day Cheryl talked to her mother about the problem. Her mother suggested, "Why not find a way to reward them, instead of harassing them?" Cheryl came up with the idea of Hershey kisses—small, cheap, and easily distributed. Workers got a kiss when they did fill out their time cards by the end of the week. All of a sudden people went crazy beating the deadline. They figured they had to fill the time card in anyway, so why not do it on time and get the kiss? Workers on travel would call in: "My plane is delayed, but I'll have it done Monday by 10:00 A.M. Will that be okay?" Cheryl usually granted them dispensation. The whole dynamic around time cards changed.

By offering an appropriate incentive, Cheryl gained our compliance. I call it the VIP Treatment, for an incentive that is visible, immediate, and personal. Cheryl had been trapped by the assumption that she had to cajole and threaten. Once freed from that assumption, she could find the right incentive to get us to do her bidding.

Notice that creative desperation is more conscious and deliberate than spotting connections, coincidences, curiosities, and contradictions. People aren't accidentally stumbling onto insights. They are actively searching for them.

Another example of creative desperation came from a conversation I had with a friend of mine, David Charlton. When he was a senior manager at Corning Incorporated, a world leader in glass and ceramics, Charlton reviewed a concept for a new way to create reactor vessels for processing chemicals that produce heat in reactions. A group of chemical engineers had criticized the materials used in this reactor, arguing that it wouldn't work because of heat transfer. The new material was a fusion of two different kinds of glasses, and the chemical engineers warned that the data were clear: the heat transfer properties of the glass would result in poor dissipation of high heat during the chemical reactions.

Charlton thought the project had a lot of merit so he tried to find some way around the objections of the chemical engineers. He noticed an unusual feature about this product—the glass walls of the reactor were very thin and the internal volumes were very small—orders of magnitude thinner and smaller—compared to conventional reactors. Perhaps the historical experience on heat transfer in large volume reactors wouldn't apply to glass reactors this thin and small.

Charlton was right—the historical experiences were misleading. New tests showed that the heat transfer issue wasn't a showstopper after all. Charlton's insight, like the others in this chapter, was to find an assumption that was trapping him and could be overturned.

NAPOLEON FINDS A PRESSURE POINT

In 1793, Napoleon Bonaparte faced the issue of how to attack a much stronger Anglo-Spanish force occupying the city of Toulon. The fate of the new French Republic depended on repelling the invaders, but the French military commander at Toulon was stymied.

Toulon, a port city on the Mediterranean, was the primary naval arsenal of the French Republic and its main southern port. Following the French Revolution in 1789, the French Republic struggled to consolidate its authority. Insurrections broke out in a number of French cities, including Toulon. The opponents of the Republic sought to restore the monarchy. The Republican forces had driven out the royalist insurrectionists in Toulon, only to be beaten back by a stronger royalist force that occupied Toulon and requested help from outside forces.

On August 28, 1793, an Anglo-Spanish fleet landed 13,000 troops to defend Toulon. It took control of the great naval arsenal and captured almost half the French navy. Unless the Republicans could repel that force, they'd have to concede the control of the seas to the British. Worse, royalists in other French areas would be encouraged to join the revolt. Already the royalists in Toulon swore allegiance to a young Louis XVII as the new King of France.

The French Republic desperately needed to drive out the Anglo-Spanish occupiers. The Republican army had besieged Toulon, but without effect because the Anglo-Spanish forces were resupplied by sea. The invaders were too strong, too numerous, and too well-defended to be defeated by force.

Napoleon relied on creative desperation to invent a new tactic. This victory in 1793 moved Napoleon from obscurity to fame and power. Napoleon had arrived at Toulon as an artillery captain. He left as a Brigadier General. He was only 24 years old.

He found a central assumption he could jettison: He didn't have to overpower the invaders, or force them to surrender. He didn't even have to attack them. He just needed to get them to leave Toulon. And he could do that by disrupting their resupply route. In studying contour maps, Napoleon noticed a lightly guarded fort, Fort Mulgrave, on Little Gibraltar, a hill overlooking the Toulon harbor. Ft. Mulgrave had no value for directly attacking the invaders in Toulon, but Napoleon saw how he could use the fort to control movement in and out of the harbor and prevent the invaders from getting supplies.

Ft. Mulgrave didn't have enough firepower to threaten resupply ships, but by capturing it from the British Napoleon would threaten two larger forts at the base of Little Gibraltar, right at the mouth of the harbor, and could use their heavy cannon to control shipping in and out of Toulon. And he could use light artillery to capture Ft. Mulgrave.

Light artillery was a class of weapon that had been developed 10 years earlier. Unlike the cumbersome heavy cannons that were the mainstay of artillery power, light artillery was much easier to transport. Animals (or soldiers) could roll it up Little Gibraltar to take Ft. Mulgrave from the British. Napoleon had made a study of the capabilities of light artillery. Remember—he was an artillery captain. His attention was drawn to Little Gibraltar because he was looking at the battlefield through the eyes of a light artillery enthusiast.

Just as Admirals Yamamoto and Stark saw the implications of the Battle of Taranto for Pearl Harbor, Napoleon was aware of the American tactics during the Revolutionary War for independence from Britain. To lift the siege of Boston in 1776, Henry Knox, an artillery officer with George Washington's Continental Army, dragged light cannons up Dorchester Heights. When the British down below realized these cannons could cut them off from their navy and from reliable supplies, they quickly sailed away. In 1781, the same thing happened at the siege of Yorktown.

The French navy cut off the British troops in town from their navy at sea and the British army surrendered to George Washington, ending the war. Napoleon knew that the besieged British troops at Toulon would worry about getting separated from their naval resupply.

The commander of the French Republic forces, Jean François Carteaux, laughed at Napoleon's plan. Carteaux was trapped by the assumption that his army would have to capture Toulon and force the invaders to surrender. That assumption anchored his war plans and he was unable to let go of it. He didn't have Napoleon's flexible mind.

It wasn't until Carteaux was fired because of his failure to dislodge the invaders and replaced by a more sympathetic general that Napoleon got the chance to use his tactic. It worked just as he expected (although he was wounded in the thigh by a bayonet during the assault on the hill). Once Napoleon took control of Ft. Mulgrave, the British commander at Toulon quickly sailed away with all the invading army. Napoleon had triumphed and was on his way as the one of the greatest tactical military commanders of all time. He went on to win 50 out of 54 battles, often against larger armies, until his defeat at Waterloo.[i]

[i] Duggan, W. (2007). *Strategic intuition: The creative spark in human achievement.* Columbia University Press.

CHAPTER SEVEN

Different Ways to Look at Insight

MY ATTEMPT TO PENETRATE THE MYSTERY of how insights originate had turned up five candidates: connection, coincidence, curiosity, contradictions, and creative desperation. Each candidate had some appeal. The connection theme was the most common, but the creative desperation theme matched the way scientific researchers thought about insight. To sort things out, I tried several different types of investigation. I studied the coding results for the data I had collected. I reviewed the scientific literature on insight. And I dived into the stories themselves.

LOOKING AT THE DATA

The first approach was to review all the data Andrea Jarosz and I had coded. The five candidate explanations of insights described in the previous chapters all came through. I found connection insights in 82 percent of the cases. Contradictions showed up in 38 percent of the cases. Coincidences played a role in 10 percent of the cases. Curiosities contributed to 7½ percent. Impasses and creative desperation were found in 25 percent of the cases. As you can see, the total for all five adds up to more than 100 percent because some of the cases coded for more than one of the themes. They weren't mutually exclusive.

Unfortunately, the data did a better job of demolishing ideas than of building them. For example, insights are supposed to depend on having an open mind, but two-thirds of the contradiction insights stemmed from a suspicious mind-set, not an open one.

Graham Wallas's advice to engage in specific preparation didn't hold up very well either. While most of the insights in the sample were deliberate, 98 out of the 120 cases, 18 percent were accidental, such as Jocelyn Bell Burnell discovering pulsars when she was looking for quasars, or Martin Chalfie hearing about the green fluorescent protein. These accidental insights couldn't rely on a preparation stage because the person wasn't wrestling with a problem.

Wallas's second stage, incubation, ran into even more serious trouble. Only 5 of the 120 cases in my sample clearly involved incubation. In 39 percent, incubation was impossible. For instance, Wagner Dodge running for his life from a raging wildfire didn't have a chance to engage in incubation. For the remaining cases, more than half, the incident accounts weren't sufficiently detailed to determine what was happening. Regardless of how we count these, incubation doesn't seem to be necessary for insights.

Insights are supposed to pop into our heads without any warning, flashes of illumination that send us running into the streets shouting, "Eureka!" One instant we are baffled, and the next instant we understand everything. Darwin had an "aha" experience after reading Malthus. Chalfie certainly had it when he listened to the lunchtime lecture about jellyfish that glowed.

Most of the cases I collected, 56 percent, were sudden, "aha" types of insights, but the other 44 percent were gradual. The notion of gradual insights seems bizarre to many insight researchers. To them, the sudden "aha" experience is what sets insight apart. It is a defining feature. These scientists object to the very idea of gradual insights.

I suspect that some insight researchers have gotten sidetracked by the "aha" experience and have lost sight of the phenomenon they set out to study. As I argued earlier, I don't believe insights are the same as "aha," any more than conception is the same as orgasm. Research is easier to do on the visible phenomenon, the "aha," but my quest was to understand insight, not facilitate experiments.

How can insights be gradual? One way is through coincidences. When Michael Gottlieb and his colleagues encountered their first AIDS patient, they were puzzled. By the fifth patient, they were ready to write up their findings on a new, mysterious, and terrifying disease. Someplace among these five patients, they spotted a repetition of symptoms that formed the kernel of their finding.

Coincidences mounted as a gradual swelling of suspicion for yellow fever, cholera, ulcers, and AIDS. There was no single event that made the Denver coaches realize they had to stop LeRoy Butler, the linebacker for the Green Bay Packers. That insight came only after seeing him disrupt play after play. Jocelyn Bell Burnell didn't suddenly glom on to pulsars. "By about the fourth or fifth time I came across this signal, my brain said, 'You've seen something like this before.'"

A second route to gradual insights is an incremental process, one small breakthrough after another. Revolutionary technologies such as Johannes Gutenberg's development of the printing press or Ford Motor's development of mass production depended on solving several problems, making steady progress rather than having a single instant of inspiration. Steven Johnson writes about slow hunches in innovation, in contrast to the flash of illumination. My data on gradual insights certainly fit the notion of slow hunches. Even when it happens, the "aha" experience, everything finally snapping into place, marks the culmination of the insight process. It isn't the insight itself.

Gradual insights also emerge from deliberate searches such as the one Meredith Whitney conducted to see if Bear Stearns was in financial trouble. As she uncovered more and more evidence, she came to realize that the firm was unlikely to recover. This is a third route.

Here is an example of a deliberate insight process that I went through to understand the surprising end to a 2007 boxing match. I had to replay the sequence many times, adding a bit more understanding each time, before I came to see how Floyd Mayweather knocked out his opponent.

THE LUCKY PUNCH

In December 2007, I watched a televised broadcast of a boxing match between Ricky Hatton and Floyd Mayweather Jr. for the welterweight boxing championship. Both fighters were undefeated. Hatton, the junior welterweight champion, was an aggressive fighter, continually swarming and pressuring his opponents. His record going into the fight was 43–0, with 31 knockouts. Mayweather, the welterweight champion, had a record of 38–0, with 24 knockouts. He was known for his defensive skills and pinpoint punching accuracy rather than for his power. At the time of the bout, ring experts considered him pound-for-pound the best fighter in the world—more dominant over the contenders in his weight class than any champion in any other division.

In the tenth round of a twelve-round bout, Mayweather caught Hatton with a lucky punch, a left hook, and knocked him down. Hatton managed to get up but was so woozy the referee stopped the fight, which had been close up to that point.

I wondered about that lucky punch and decided to study it more carefully. I obtained a video of the fight and replayed the knockdown.

What I saw surprised me. Hatton didn't simply walk into a punch. Watching it in slow motion, I saw that Hatton was throwing his own left hook and Mayweather just beat him to the punch.

I played the sequence again. Hatton moves in. Mayweather is trapped in his own corner of the ring. Hatton throws his left hook. Mayweather is just standing there. But as Hatton throws his punch, he drops his right hand, giving Mayweather the opening he needs.

I rewind and play it again. I watch Hatton throw his left hook, and I see that he is about halfway into the punch before Mayweather even moves. Mayweather finally puts his right glove up to block Hatton's punch and brings his own left hand *back* to get leverage for his punch. He starts bringing his left fist back even as Hatton is halfway into his punch. Then Mayweather unleashes his left hook.

So I was wrong about the lucky punch. Mayweather deliberately waited for Hatton and then started his punch when he saw what Hatton was doing. Hatton was throwing a big, slow, sweeping left hook, and Mayweather clipped him with a "cut" left hook, a shorter punch with a small windup and a shorter, faster arc. He used Hatton's momentum to magnify the force of the punch.

Where did that come from? Mayweather and his trainer must have studied tapes of Hatton to see his tendencies, such as his habit of dropping his right hand when throwing a left hook, but Mayweather would have needed to get the feel of those tendencies in the ring. Back to the rewind button.

Hatton doesn't throw his big left hook at all in the first round. It makes its first appearance in the second round, four times, each the same. The fighters are separated; Hatton bounces forward, telegraphing that he is going to throw a punch with his left hand. Hatton's right hand isn't a threat; it isn't even a shield because Hatton drops the right glove as he puts force into his left. Each time Hatton throws his left hook, his power punch, Mayweather just ducks or blocks the punch, taking its measure. At other times Hatton fakes the left hook and instead throws a straight left jab. Mayweather is going to have to wait a fraction of a second to see if Hatton intends to jab or hook.

In round three, Hatton throws his big left hook only once. Mayweather has started to time it, throwing his cut left hook for the first time to beat Hatton to the punch but not doing much damage. They are in Mayweather's corner. Throughout the fight Hatton throws that left hook eighteen times (that I could count); ten of them are in Mayweather's corner of the ring, the corner where Hatton eventually gets flattened. The only times Mayweather responds with his cut left hook are in that same corner. It is the place where Hatton prefers to launch his left hook attack. Perhaps Mayweather retreats there because he feels safe, choosing it for his ambush. It is the kill zone.

During the eighth round, in a premonition of the end, Hatton traps Mayweather in Mayweather's own corner, throws the left hook, and gets nailed with Mayweather's cut left hook. Hatton is stunned but recovers. If I didn't know better, I would assume I was seeing the knockout punch two rounds later.

After getting tagged in the eighth, Hatton doesn't use his left hook in the ninth round, but the fight is slipping away from him. By the tenth round perhaps Hatton is getting tired or desperate. He starts winding up with his left hook, four times in short succession at five seconds, seventeen seconds, and twenty-four seconds into the round. The fourth time, sixty-one seconds into round ten, is the final installment. Hatton gets clobbered, and so does my conviction about the lucky punch.

The fight illustrates to me how Mayweather learned Hatton's tendencies and tip-offs, how he learned to time Hatton's left hook. Many years ago when he was the heavyweight champion of the world, Floyd Patterson commented that boxing was like chess but with bodies instead of pieces. After reviewing the Mayweather-Hatton fight, I had a better idea of what Patterson meant. My insights about Mayweather's strategy came about gradually. There was no "aha" instant when it all clicked into place.

My review of the coded data raised doubts about several popularly held beliefs about insights but didn't provide any answers. I still didn't understand the link among the five categories of connections, coincidences, curiosities, contradictions, and creative desperation. To sort that out, and hoping to acquire additional ideas about insights, I switched to a second type of investigation.

LOOKING AT THE SCIENTIFIC LITERATURE

When I began this project, I carefully collected more than eighty recently published scientific research papers on insight, plus about fifteen books. Then I wondered if I should read them. My training is to start any research study by reviewing the scientific literature, but I wanted to have fresh eyes in gathering and examining all the cases of insight I had collected. If I reviewed the scientific literature, I would be seeing all these cases through the same lenses as other investigators. I would interfere with my own process of discovery.

Therefore, I put them all aside and didn't read any. It was a risky decision. I might have missed some important issues, gone down blind alleys, or wasted my time reinventing what other researchers already knew. I accepted these risks because I wanted to gain my own impressions and build my own mental model of insight. I did *not* want to have a prepared mind that might get in the way of finding something new.

Nevertheless, really good researchers and theoreticians have pondered questions about insight in the past; I had a professional responsibility to study their ideas. Also, I was curious about what they had found and about their perspectives. After I had collected and categorized most of the cases, I judged that I was far enough along to check out other views. I retrieved those articles and books and worked my way through them. One of my favorites was *The Nature of Insight,* a 1995 collection of chapters from leading insight researchers. I found a lot of useful ideas in that book and the other sources, but also a lot of confusion and disjointedness that mirrored my own impressions. The different researchers argued with each other about all kinds of fundamental issues. Some claimed that the "aha" experience was an essential part of insight and that cognitive restructuring wasn't necessary. Others argued just the reverse. Some of the researchers suggested that people don't even rely on insight processes to solve insight problems and that the concept of insight is useless. After sifting through the scientific papers, I was glad that I had not gotten drawn into these debates.

My literature review included some of the research on decision biases to see how they lined up with insights. Decision biases are systematic deviations from rational standards of good judgment. For example, we might judge that Americans are more likely to die in a car crash than from stroke, probably because TV and newspapers often report car crashes. We look at the dramatic photographs of twisted metal and think that there's no way we could have survived. There are fewer photographs and stories about stroke victims even though stroke causes four times as many deaths as traffic accidents. Therefore, our estimate is biased by what we see in the media. I speculated that in some ways, insights are a counterpoint to decision biases.

In the early 1970s, the field of psychology produced a new discipline devoted to studying decision biases. Two Israeli researchers, Danny

Kahneman and Amos Tversky, founded the heuristics-and-biases movement, which catalogs and explains the different kinds of biases afflicting us. Kahneman won a Nobel Prize for this research (Tversky had died before he could share in this honor). The heuristics-and-biases movement has given rise to the discipline of behavioral economics, uniting economists and psychologists in a quest for ways to exploit decision biases for socially valuable purposes.

In his recent bestseller *Thinking, Fast and Slow,* Kahneman distinguishes between fast and intuitive System 1 thinking and System 2 thinking, which is slower, more critical, analytical, and deliberate. The heuristics-and-biases movement has shown that decision biases largely stem from System 1, the cognitive impulses we have. System 2 thinking refers to our mental strategies for monitoring these impulses to inhibit and correct them when necessary. The heuristics-and-biases community advocates methods for strengthening System 2 so that it can do a better job of controlling System 1.

These ideas line up nicely with the two arrows in the performance diagram I presented in Chapter One. System 2 is all about the downward arrow. It is all about reducing errors. The weight of the heuristics-and-biases studies, reporting bias after bias, can unsettle us, making us doubt our own judgments, our capacity for clear thinking. Frequently, the message of the heuristics-and-biases research is that we are likely to draw mistaken conclusions unless we engage in rigorous critical thinking. The message is that we have to strengthen the down arrow.

I believe it is important to counterbalance this negative impression of System 1 with a sense of awe and appreciation about the insights we create and the discoveries we make. We need both arrows, just as we need both System 1 thinking and System 2 thinking. The process of gaining insights, the upward arrow, balances the worries about decision biases, the downward arrow. This perspective on the heuristics-and-biases research strengthened my interest in understanding the up arrow.

I had tried looking at the data and then at the literature. It was time for the third approach.

LOOKING AT THE STORIES

I had faced a similar confusion some thirty years ago. I had been evaluating different models of decision making, but none fit the data I'd collected. I had interviewed twenty-six experienced firefighters and recorded thirty-two stories of tough decisions they had made under time pressure and uncertainty. And I had gotten stuck. The stories didn't support my own expectations or the standard accounts of how people compare options to make decisions.

In desperation (I am tempted to call it creative desperation), I ditched all these hypotheses and patiently read through these stories to see what they were telling me. I put my trust in the stories, not in the theories. The result was a discovery of a different model of decision making, a recognitional model, that dispensed with the notion that people have to compare options. That definition of decision making—a comparison of options—had been trapping me, and I could only abandon it after I immersed myself in the stories.

In the years since, I have relied on methods for collecting critical incidents—stories—to understand the way people think. I have had some satisfying successes, along with some failures. Perhaps the most gratifying success is that these methods for collecting and analyzing stories have given rise to a community of naturalistic decision researchers.

Naturalistic methods can be a bit nerve-wracking because you never know what you are looking for. You sift through the stories, on the lookout for patterns that might be meaningful. When you do laboratory studies in psychology, you define in advance what data you're going to collect, what hypotheses you're going to test, what statistics you're going to use. But the story-based strategy leaves all of that open. You can't define in advance how you are going to analyze your data because you don't know what patterns might emerge. Reviewing the stories is scary and exciting.

The incident-based method is well suited for the early stages of science in which investigators try to explore some phenomenon. I hoped to find some answers in the 120 stories of insights.

The data coding seemed to cluster the cases into five insight themes, and that turned out to be a useful finding in itself. Connections, coincidences, curiosities, contradictions, and creative desperation. Which of these was the answer? Or did I have to blend them, and if so, what would that blend look like?

There were some glimmers, hints of a way forward. I was taken by the contradiction form of insight, particularly the insights of each of the financial experts who foresaw the 2007–2008 crash. I was inspired by John Snow's discovery of how cholera is transmitted. I kept returning to my admittedly feeble understanding of the Einstein example, the way he developed his theory of special relativity. The Chalfie case with the green fluorescent protein seemed significant.

I wondered if I could build on my definition of insight as an unexpected shift to a better frame—a better story—for understanding how things work. Was there any daylight there? The notion of a frame includes slots for fitting data into the frame. In each of my cases a few of the slots usually stood out. They were more important for anchoring the story than the other slots were. Would that idea give me any leverage, or was it just a set of mixed metaphors, anchors, slots, and leverage?

It was all very vague. These glimmers were just fragments of ideas. I hoped these fragments might be useful, but I didn't anticipate how they would lead to a new model of insight, an explanation of how we make discoveries.

CHAPTER EIGHT

The Logic of Discovery

I NOTICED SOMETHING ODD as I thumbed through the stories: I had five insight strategies, and two of them worked in opposite directions. They weren't just different from each other. They seemed to be reverse activities. No wonder I couldn't synthesize the strategies.

When faced with creative desperation, we try to find a weak belief that is trapping us. We want to *jettison* this belief so that we can escape from fixation and from impasse. In contrast, when using a contradiction strategy, we *center* on the weak belief. We take it seriously instead of explaining it away or trying to jettison it. We use it to rebuild our story.

Look at some of the creative desperation stories. Wagner Dodge running for his life hit upon the idea that he could get rid of the tall grass, depriving the forest fire of its fuel. His insight suggested an unconventional way to eliminate the grass. He set it on fire. Aron Ralston, trapped by the boulder, gave up his goal of saving or extricating his right arm and gave up his goal of using his knife to cut through his arm bones. Instead, he used the leverage from the boulder to snap the bones. Cheryl Cain, irritated with the people in my company who wouldn't fill out their time cards at the end of the week, gave up on threats and replaced them with kisses. Napoleon at Toulon discarded the assumption that the French would have to overpower the invading British-Spanish forces. He merely had to threaten their resupply lines to make them sail away.

Now compare these with the contradiction stories. John Snow started having doubts that cholera was spread by bad air—the miasma theory—when his Tilt! reflex was triggered by contrary evidence. Some people came down with cholera, but others didn't, even though they all breathed the same air. Snow's own autopsies showed that cholera victims had healthy lungs but ravaged digestive tracts, more evidence that contradicted the miasma theory. Snow could have explained away these anomalies. Our natural tendency is to explain away the deviant data points. And for good reason. Most of the time that deviant data point is wrong or else it doesn't really apply to our situation. It is a safer strategy to discount the anomalies, but that safe strategy doesn't yield insights. We arrive at insights when we take the anomaly seriously, as Snow did. He found a different frame, a different story. He speculated that contaminated water was spreading cholera. This perspective guided him to collect data that others ignored.

Look back at Einstein's discovery, the way he developed his theory of special relativity. For a decade Einstein engaged in thought experiments about light beams, sometimes uncovering anomalies and inconsistencies. For instance, if he rode a light beam, then he would be moving at the speed of light. From his perch on that light beam a second light beam would move away from him at the speed of light. Yet nothing can move faster than light, so Einstein had found a contradiction. The easy answer is to explain this anomaly away by insisting that the second light beam wouldn't really move away from him at the speed of light. Instead of trying to explain it away, however, Einstein took it seriously. And to do that, he had to revise some of the other beliefs, the other anchors that seemed much firmer, in particular the Newtonian view that space and time were constant. If we make the speed of light the constant, then time slows down the faster we move, lengths contract, and masses increase. We enter the world of special relativity.

The contradiction path to insights seems to follow a common script: we encounter an anomaly, we resist the temptation to discard it, we give it credence and imagine that it is valid, and then we revise the rest of our beliefs to make it all fit. Usually we have to discard some of our earlier beliefs, even major ones that have anchored our understanding up to now. We give up the notions of space and time as fixed entities. We give up miasma theory. We continue with this line of reasoning until we arrive at a new frame that has some of the old anchors, but also centers on the anomaly.

My sample had 45 contradiction insights, triggered by an anomaly or an inconsistency. The person who had the insight explored the anomaly rather than explaining it away in 42 of these cases. The other 3 incidents didn't have enough detail to assign a code. In none of the contradiction insights did the person explain away the anomaly. That's not how contradiction insights work.

The investors, Eisman, Paulson, Burry, Lippmann, and Park, seemed to follow the same path as they resisted the gravitational pull of the financial bubble. They hit a contradiction between the extreme confidence of the mortgage and Wall Street communities that housing prices would continue to rise and their own data points suggesting that the end was in sight. The investors anchored on the anomalies they uncovered, even though their conclusions were inconsistent with the dominant views on Wall Street, and they worked out the implications from there. They sensed that the contradiction was important. It was an opportunity rather than an inconvenience.

Both the contradiction cases and the cases of creative desperation directed people's attention to outliers—important beliefs that weren't firmly supported. However, in the desperation cases people did not look for ways to reframe their understanding around the weak, anomalous anchor. Just the reverse. People who are in desperate straits attack the weak anchor instead of embracing it. In times of desperation, we actively search for an assumption we can reverse. We don't seek to imagine the implications if the assumption was valid. Rather, we try to improve the situation by eliminating the assumption.

So now I had distinguished two different paths to insight.

- They spring from different motivations—wanting to escape from a bad situation versus wanting to rethink conventional wisdom.

- They have different triggers—searching for a flawed assumption versus encountering an inconsistency.
- They rely on different activities—replacing the flawed assumption versus building on the weak assumption that leads to the inconsistency.

They also have some similarities. These insights are disruptive in that they don't let us retain our comfortable beliefs. Instead we have to modify the core beliefs that anchor our understanding. We abandon some beliefs/anchors and revise others. The two paths also lead to the same kinds of outcomes. We change what we understand. In addition, we sometimes change our ideas about the actions we can take, the way we see situations, how we feel about things, and what we desire.

The diagram of a Triple Path Model of insight tries to capture these features. It is a *Triple Path* Model because the remaining strategies, connections, coincidences, and curiosities are combined in the third path, which is shown as the middle column.

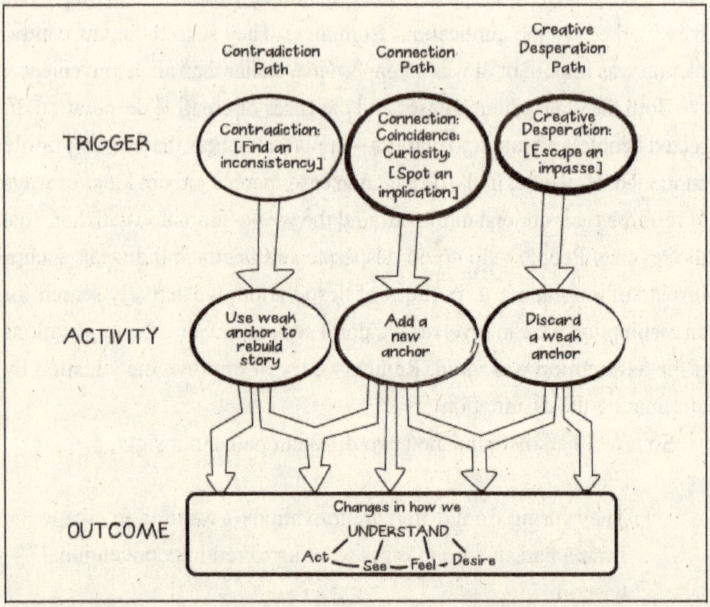

The connection path is different from the desperation path or the contradiction path. We're not attacking or building on weak anchors. When we make connections or notice coincidences or curiosities, we add a new anchor to our beliefs and then work out the implications. Usually the new anchor comes from a new piece of information we receive.

Martin Chalfie followed this path when he learned about the green fluorescent protein. He combined the idea of those proteins with his translucent worms and discovered a new research direction. The green fluorescent protein was a new anchor that Chalfie added to the mix.

Charles Darwin followed the same path upon reading Malthus's work on how populations grow to the point at which members have to compete for scarce resources such as food and water. This notion of competition for scarce resources was a new anchor that helped Darwin fit his theory of evolution together. Malthus's theory didn't contradict any of Darwin's ideas, nor was Darwin desperately searching for a breakthrough. He just appreciated what would happen if he tossed Malthus into the mix of ideas he was considering.

Both Admiral Stark and Admiral Yamamoto, an American and a Japanese, recognized the implications of the Battle of Taranto. Aircraft carriers could launch a surprise attack on a naval fleet. What had seemed a safe harbor could be an immobilizing trap. This aircraft carrier tactic added a new anchor to their frames for planning to destroy a naval force.

Alison Gopnik seized on her two-year-old son's comment that pineapple in kirsch was yucky/yummy to add a new anchor to her method for studying infant intelligence. The comment didn't contradict anything she knew; it gave her an idea for probing the way fourteen- and eighteen-month-old babies think.

Sometimes the new anchor comes from seeing a coincidence. Carlos Finlay spotted a coincidence between outbreaks of yellow fever and the *Culex* mosquito. He treated the mosquito as a new anchor, a new and critical part of the frame for explaining how yellow fever spreads.

I've combined the connections, coincidences, and curiosities in the Triple Path Model. They have the same dynamic: to build on a new potential anchor. They have the same trigger: our thinking is stimulated when we notice the new anchor. Coincidences and curiosities aren't insights in themselves; they start us on the path to identifying a new anchor that we connect to the other beliefs we hold. Connections, coincidences, and curiosities have the same activity: to combine the new anchor with others. This path to insight doesn't force us to abandon other anchors. It lets us build a new story that shifts our understanding. This path has a different motivation, a different trigger, and a different activity from the contradiction and the creative desperation paths. Nevertheless, like the other two paths, the outcome is the same: an unexpected shift in the story.

This shift isn't a minor adjustment. It means changing the core components, the anchors, used in the story. With contradictions and creative desperation, some of the previous anchors are discarded. With connections, a new anchor is added. In all the paths the anchors in the story after we make an insight are different from the ones we started with.

Each of the three paths, the contradiction path, the connection path, and the creative desperation path, gets sparked in a different way. And each operates in a different fashion: to embrace an anomaly that seems like a weak anchor in a frame, to overturn that weak anchor, or to add a new anchor. Future work on insight is likely to uncover other paths to insight besides the three shown in the diagram.

The bottom of the diagram of the Triple Path Model shows different routes to the outcome. It shows some direct routes from the three strategies, but also some crossovers. That's because many of the insights in my sample relied on more than one path. For example, Carlos Finlay noticed a coincidence between *Culex* mosquitoes and yellow fever. Then, with the mosquito hypothesis in a dormant stage, Walter Reed spotted a contradiction between miasma theory and yellow fever: the instance where one convict died and the others lived even though they shared the same air, clothes, and bedding. Finally, Walter Reed's team encountered a new anchor: the incubation period inside the mosquitoes. When they connected an incubation period to the mix, they showed that mosquitoes do spread yellow fever.

Multiple paths are the rule, not the exception. The diagram doesn't show all possible blends, which would get complicated. Most of the blends relied at some point on the connection path, and so I have portrayed the other two paths joining the connection path. There were only a few cases that drew on both the contradiction path and the creative desperation path.

With the Triple Path Model, I feel that I've solved the mystery behind the flash of illumination. Of course, that isn't a mystery that can be solved. It doesn't have a "right" answer. It isn't like John Snow discovering that cholera is spread by contaminated water or Harry Markopolos concluding that Bernie Madoff was doing something illegal. Those questions do have a right answer that can be verified.

Nevertheless, I accomplished more than I expected. I'm satisfied that I have a fuller picture of how insights work. The best we can do when faced with challenges like explaining the origins of insight is to move the posts forward, and I believe I've done that. I believe I've moved beyond Graham Wallas's four-stage account of preparation, incubation, illumination, and verification, and beyond any of the other accounts of insight that I'd read about. I did it by puzzling over the stories, not by grinding out the data analysis.

While the Triple Path Model isn't the last word on insight, I see it as the next word. It seems like a plausible answer to the question about how the up arrow in the performance equation works. It feels like a richer, more comprehensive, and more useful account. In Part III, we'll examine what the Triple Path Model says about increasing insights.

The Triple Path Model shows why earlier accounts aren't wrong as much as they are incomplete. They restrict themselves to a single path. Researchers and theorists such as Wallas who describe insight as escaping from fixation and impasses are referring to the creative desperation path. Researchers who emphasize seeing associations and combinations of ideas are referring to the connection path. Researchers who describe insight as reformulating the problem or restructuring how people think have gravitated to the contradiction path. None of them are wrong. The Triple

Path Model of insight illustrates why people seem to be talking past each other. It's because they're on different paths.

Now that we have a fuller picture of how insights work, it's time to turn to the dark side of insights, to the forces that stifle the discovery process. It's time to turn to the second mystery that captured my attention. What stops people from having more insights?

PART II

SHUTTING THE GATES

What Interferes with Insights?

CHAPTER NINE

Stupidity

WHEN I STARTED INVESTIGATING why people fail to have insights, I remembered an evening in late September 2003 when I was involved in the heat of a fantasy baseball pennant race. If you are not involved in one of these leagues, fantasy baseball may seem trivial or nonserious. However, those who do play appreciate the passion that drives us to beat the other teams and the mental energy and time we put into making our decisions. Competitors act as managers of teams whose performance depends on the accomplishments of actual baseball players. Each "manager" assembles a team made up of active baseball players and racks up points depending on how well those players perform in real games.

As a manager, I might have ten pitchers on my team. If they win a lot of games and strike out a lot of batters, my scores go up. If my pitchers get clobbered, then my scores go down. The season lasts as long as the regular baseball season, from early April to the end of September. The fantasy managers jockey all season long to come out on top at the end. The managers are scored using categories such as how many games the team's pitchers won and how many overall strikeouts they had.

My daughter Devorah joined the league in 2003, playing remotely via computer. It was her rookie season, and she was hoping not to disgrace herself. She was aware of all the trash talk and the public humiliation of people who make stupid trades. She didn't do badly—she was in seventh place out of ten teams with a chance to move into fifth at the end of the season. The outcome depended on Devorah's pitchers and whether they would win enough games and strike out enough opposing players. You might think that the difference between fifth and seventh place isn't a big deal, but once the competitive fever takes hold, fantasy baseball players start obsessing about every edge they can get.

Devorah lives in the Boston area, but on Saturday evening, September 27, 2003, she was visiting me at my home in Yellow Springs, Ohio, so I got to watch her in action. She was worried about her pitchers. Our league had set a maximum of 1,250 innings a team's pitching staff could compile over the entire season. After a team went over this maximum, none of the statistics for any of its pitchers would count.

Devorah was very close to this maximum. In fact, Devorah's team had compiled 1,249 innings over the season and had only one inning left going into Sunday, the last day of the season. (A baseball game lasts for nine innings unless the score is tied, and then the game continues for extra innings until there is a winner.) Devorah had one pitcher scheduled to start on Sunday. She needed every win and every strikeout he could get. She asked me if her team would get credit only for any of her pitcher's strikeouts during the first inning, or would the team get credit for all the strikeouts he recorded that game? And would she get credit for a win if her statistics shut down after the first inning?

I commiserated with her—it must be frustrating not to know how the computer program calculated these things. Devorah suggested that we look up the rules. Even though I had been playing fantasy baseball for many years, I wasn't familiar with the new computer-based version we were now using. I didn't know I could look up the rules or even where to find the rules.

Devorah located the rules button on the home page, clicked on it, and found out that all pitchers' statistics counted on the day that the team exceeded its maximum innings. After that day, no pitcher's statistics would be counted. So we had the answer: Devorah's team would get credit for all of her pitcher's strikeouts on Sunday, not just the strikeouts he scored in the first inning. And if he won the game, she would get credit for that as well.

"Okay," I said, "you don't have to worry about tomorrow." I was satisfied that we had resolved the question. But I noticed that Devorah wasn't really listening to me. She was thinking about something else.

"You know, it's too bad I only have one pitcher in action tomorrow because all my pitcher innings will count, and tomorrow is the last day of the season. But there's nothing preventing me from picking up pitchers who are scheduled for tomorrow. I have no time to lose."

Devorah's insight had two parts, understanding and action. First, she didn't have to worry that her pitching staff would have too many innings. That was her new understanding. Her original question was about her one pitcher who was scheduled to appear, but she saw that her entire pitching staff was free from the 1,250 innings limit. Second, her understanding translated into action. She could replace pitchers on her team to get as many as possible who were slated to play on Sunday. She could drop from her team all the inactive pitchers (starting pitchers play only every fourth or fifth day) and replace them with free agents—pitchers who hadn't been claimed by any teams in our fantasy league and were scheduled to play on Sunday. All their statistics would count even though Devorah's team would go way over the 1,250-inning maximum. It was like being at an all-you-can-eat buffet.

I hadn't noticed that loophole. No one in our league history had ever done such a thing. It was 11:00 P.M. The deadline for picking up unclaimed players was 12:00 midnight.

Devorah rapidly dropped all the pitchers who weren't starting and replaced them with ones who were scheduled to pitch on Sunday. "I'm dropping Roger Clemens [a pitching superstar] for Matt Clement [a journeyman]," she exclaimed disbelievingly.

On Sunday, Devorah picked up the points she needed because her new set of pitchers scored several wins and collectively racked up lots of strikeouts, enabling her to slip into fifth place.

Devorah's actions were within our ground rules; only later did we find that her tactic was well-known in other leagues. But Devorah independently invented it. I was there. I watched her read the rule, see that it answered her question, and then make an unexpected discovery about how she could reshape her entire pitching staff.

You may look at what Devorah did and conclude that this is nothing special—this sort of insight happens all the time. Exactly. We all continually have insights, mostly small and not terribly significant.

This incident isn't just about Devorah's insight. It's also about the fact that I was so clueless. I was sitting right next to her. I had the same information. Yet I was a chucklehead and missed the insight.

What happens to us to make us so stupid? What blinds us to the insights that dangle right in front of our faces? Even as I was resolving the first mystery—the paths we take to gain illumination—I became gripped by this second mystery, What prevents us from having insights?

STUPIDITY IN ACTION

Many years ago, I took a vacation with my two young daughters, Devorah and Rebecca. We flew from Dayton, Ohio, to New York City. At the Dayton airport I followed my usual travel routine and put my keys in my briefcase once I parked my car. During the trip my youngest daughter, Rebecca, developed an ear infection. We went to a physician in New York, who warned against letting her fly for a few weeks.

To return to Ohio at the end of the week, we decided to take a train. Because southwest Ohio doesn't have any train service, we took the train from New York to Toledo, Ohio. My mother-in-law, Bessie, lived in Detroit and had already planned to drive south to our home in Yellow Springs for a family event, which was why we had to get back to Ohio. Bessie was happy to pick us up in Toledo, which was on her route south. She then drove us to the Dayton airport so that I could get my car. So far so good. My replanning was working out well.

At the Dayton airport, as I got out of her car, I wondered, "Do I need my suitcase and briefcase?" I couldn't see why—I was just going to drive home and would meet Bessie there. So I left them in Bessie's car. As I approached my car, I realized my mistake.

Before that moment, I could have correctly answered each of these questions: Where are your car keys? (in my briefcase, as usual for a trip), Where is your briefcase? (in Bessie's car), and Where is Bessie's car? (driving out of the airport). I knew all the pieces but failed to put them together in time to notice the inconsistency.

I had to rent a car at the airport and return it the next day. It was an extra expense and an inconvenience, but not a tragedy.

I got off easy compared to a professional colleague who came home, got the mail, and decided to dash upstairs on a quick errand. She left the mail, including some glossy material, on the bottom stair to be retrieved and properly stored after she finished her errand. While upstairs, she decided to check e-mail, with one thing leading to another for the next several hours. It was now dark and time for dinner. She came downstairs but didn't bother turning on the light—why waste energy? After all, she knew the staircase well. Except that she had forgotten the mail. She could have answered the questions "Where is your mail?" (on the bottom step), "Does your mail provide a safe amount of friction?" (no—there are glossy brochures), and "Can you see the mail?" (no—it is dark). When she got to the bottom step, she hit the glossies and went flying. The result was a compound fracture of her leg, traction, and sixteen weeks in bed.

These are both examples of stupidity. If I had remembered that my car keys were in my briefcase, I wouldn't claim the memory as an insight. If my colleague had remembered the glossies on the stairs, no insight. Neither of us noticed the contradiction lurking within our beliefs. We both knew that our current situation differed from what we originally had intended, but we failed to see the implication.

I can also think of times when I've failed to make obvious connections. I have been wanting to show my wife, Helen, a small locker I recently rented to store file boxes and other materials, but we've never had the time. Then one day Helen and I had to run an unanticipated errand that took us right past the facility. It wasn't until we were driving back that I realized I should have grabbed my keys to the storage locker on the way out of the house. I missed the opportunity. Dumb.

These are the kinds of connections we expect to make, the kinds of contradictions and inconsistencies we expect to catch. We shouldn't give ourselves credit for catching them. The failure to catch them is what merits our attention.

These cases of stupidity seem like the opposite of insight, the opposite end of the contradiction and connection insight paths we explored earlier. When we spot really subtle inconsistencies and contradictions, we deserve credit for insight. When we fail to spot obvious connections and contradictions, we castigate ourselves. In the fantasy baseball example, Devorah and I received additional information: the rule that all of her pitchers' innings would count. She made the connection, saw the implication, and reshaped her team. I didn't make any connection with that information. The incident in which I failed to grab the storage locker keys also shows a failure to make a connection. Here the new information was the unexpected errand. I missed the implication until we drove past the storage facility.

In the example of my misplaced car keys, there was a glaring contradiction between my beliefs. I knew where my keys were, I knew I would need my keys once I reached my car, but the Tilt! reaction didn't kick in until it was too late. Ditto my friend who knew where she'd left the circulars and knew she was approaching the bottom step, but no Tilt! reaction until she literally tilted into the air.

Often we blame our memory. I should have remembered that I put my car keys in my briefcase. Actually, I sort of remembered. I could have answered the question correctly if prompted. And I spontaneously remembered as I got closer to my car and started reaching in my pocket. But the key location didn't pop into my consciousness when I needed it. Memory isn't just about calling up answers to information requests, such as the first time I saw Tom Cruise in a movie (answer: *Risky Business*). We depend on memory to spotlight information without being asked. Memory is necessary for this kind of spotlighting, but it isn't sufficient. Some more active process must be at work here, a process of flagging the right connections and contradictions to spotlight.

Examples of stupidity put actual insights into perspective. They suggest that we often engage in the insight strategies even for everyday activities that don't count as insights. We continually make connections, look for implications, spot inconsistencies, and challenge weak assumptions. When we're on automatic pilot, and the connections and contradictions are obvious, we don't give ourselves credit for noticing them.

We do give ourselves discredit for missing them. When we fail to make obvious connections, when we miss obvious anomalies and inconsistencies, when we get hung up on assumptions that are clearly wrong, we are guilty of stupidity. Perhaps each insight pathway in the Triple Path Model could be treated as a continuum from stupidity to insight, with normal alertness in between. Stupidity and insight would then be two bookends—the two poles of each continuum.

In assembling my 120 insight cases, I encountered more than 20 cases of stupidity, which I put in a separate file. I had noticed a few and then started adding others, building a new stack. The stupidity file included "false insights"—cases in which people erroneously believed they had achieved an insight and had full confidence in their cleverness. Financial bubbles often depend on false insights. The investors bidding prices up to ridiculous levels aren't just caught up in a frenzy of greed. They often delude themselves into thinking that a transformational change is in the works. The Japanese stock and real estate bubble of the late 1980s was thought to reflect Japan's growing dominance of the world economy. The U.S. dot-com bubble in the late 1990s apparently heralded the emergence of information technology that was going to result in a new paradigm of cascading productivity. The U.S. housing bubble in the early twenty-first century supposedly reflected more powerful analytical tools for managing risk. The world of financial investment is not lacking in false insights.

The topic of stupidity deserves its own treatment. However, I don't have the ambition to be a researcher of stupidity. It is bad enough to be a practitioner.

Besides, I didn't want to become distracted from my effort to understand what interferes with our ability to gain insights. Stupidity might be one reason, but there are others. Upon hearing of Darwin's theory of evolution, T. H. Huxley commented, "How extremely stupid not to have thought of that."

People like Huxley are not stupid; they just feel stupid when they miss an insight that is right in front of them. What gets in their way?

If I had studied insight using the puzzles that are the staple of laboratory research—the water jars, the nine dots, and the pendulum puzzle—I could have tested lots of college undergraduates on the same task and contrasted their successes and their failures. That's one of the advantages of using standardized puzzles. They let a researcher delve into the reasons that some people solve the problems quickly and others struggle. Unfortunately, I couldn't make systematic comparisons using the grab bag of examples I had collected.

Or could I? Perhaps there was a way to do a natural experiment.

CHAPTER TEN

The Study of Contrasting Twins

To investigate my new mystery, the reason that people fail to have an insight even if they have all the necessary information, I could take advantage of a natural comparison. Some of my 120 cases described a person who didn't have the insight along with the person who did. In fact, as I sorted through the stories again, I counted 30 such cases. I called these "contrasting twins" even though they weren't related at all, except in one way: the two "twins" were identical in having the same relevant information needed for the insight. The inspired twin pressed forward to gain the insight; the blinkered twin missed it.

My selection rule was that the twin who missed the insight had to be a real person in the story, not just an anonymous bystander. Even if the person wasn't named, I needed at least a little detail and background information on the twin who missed the insight so that I could speculate about what got in his or her way.

You have already met some of the twins in the previous chapters. Wagner Dodge dreamed up the strategy for setting an escape fire in Mann Gulch. At an official inquiry following the tragedy, the two other survivors, Robert Sallee and Walter Rumsey, testified that they never thought of an escape fire and didn't listen to Dodge when he told them to follow him into the ashes of that fire. Napoleon looked at the maps of the military picture at Toulon and formulated a strategy that his commanding officer, General Jean François

Carteaux, didn't appreciate. John Snow doubted that miasma theory explained the spread of cholera because so many data points contradicted it, while Edwin Chadwick, the London sanitation commissioner, and William Farr, the leading demographer of the time, retained their faith in miasma theory to explain cholera. David Charlton spotted a way around the data on heat transfer, unlike the materials engineers on his team who relied on those data. And don't forget how my daughter Devorah realized the implication of the fantasy baseball rule about what happens when a team's pitchers run out of innings. I was the dumb twin there, sitting next to her, oblivious to her discovery.

I sorted through the contrasting twin stories several times, trying to find what distinguished the successes from the failures. Eventually, I arrived at four reasons that we might miss the chance to have an insight: flawed beliefs, lack of experience, a passive stance, and a concrete reasoning style.

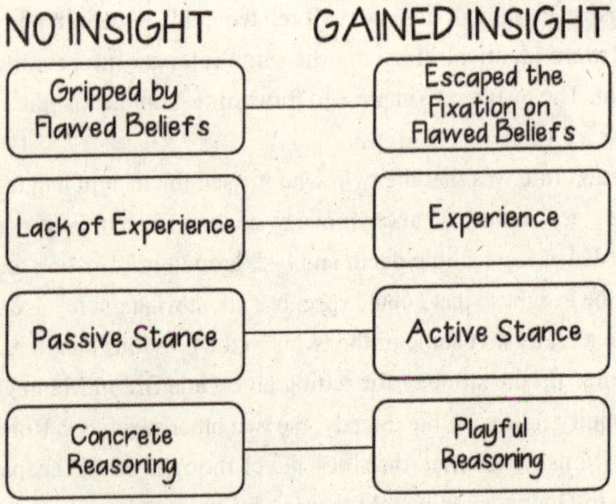

FLAWED BELIEFS

The first reason we miss insights is that we are gripped by a flawed belief. In most of the cases, 21 out of 30, the failure twins fixated on some erroneous ideas that blinded them to the discovery. Look back at the way the medical establishment accepted miasma theory for the spread of cholera. Researchers were impervious to contrary data and ridiculed John Snow for speculating about contaminated water. In these 21 cases the insightful twin was open to new data that the failure twin rejected. Sometimes the insightful twin started out with an incorrect belief but then rejected it. So the strategy for the successful twins seemed to be to speculate and test, whereas the failure twins clung to the flawed belief and wouldn't let go of it.

Here is another example, a matched pair of twins from the Cuban missile crisis of October 1962. John McCone, head of the CIA, warned that the Russians seemed to be getting ready to put ballistic missiles into Cuba, ninety miles off the coast of Florida. These missiles were capable of carrying nuclear warheads. Sherman Kent, the chief CIA analyst, strongly disagreed.

Kent's theory ran as follows. America had such strong military superiority over the Soviet Union that it seemed implausible for the USSR to do anything that might provoke a major nuclear conflict. The Americans could absorb a first-strike nuclear attack and still be able to destroy the USSR. ("Absorb" means enduring despite the annihilation of New York City, Washington, DC, and other metropolises, not an entirely comforting strategy.) Kent's study of historical trends convinced him that the Soviets would never do something as provocative, risky, and stupid as putting nuclear missiles into Cuba.

In early October 1962, when the first photographs from a U-2 spy plane showed suspicious activity in Cuba, McCone went on full alert. He had already been trying to keep track of the weapons the Soviets were pouring into Cuba following the Bay of Pigs incident in which the United States launched a futile attempt to overthrow Fidel Castro. After the Cubans foiled this plot, the Soviet Union sent sophisticated weapons to Cuba to provide protection from future invasions. McCone then worried about what *kinds* of weapons the USSR was placing in Cuba.

McCone looked at the U-2 photographs of several Cuban installations, and they seemed to be some of the most modern surface-to-air anti-aircraft missiles (SAMs) the USSR had. The SAMs in the Soviet Union were designed to shoot down prying U-2 aircraft flying at very high altitudes. Now these SAMs were showing up in several locations around Cuba. Why? The USSR must have something valuable to hide if it was going to the effort of stationing surface-to-air missiles all around Cuba. It looked to McCone as if the Soviets were getting ready to install MRBMs—medium-range ballistic missiles—and were waiting to complete the defenses before bringing in the ballistic missiles themselves.

McCone made sure that the United States intensified its surveillance of Cuba. He kept checking, even during his honeymoon in France. History has proven him correct. McCone sounded the early warning that allowed President John F. Kennedy to confront the Soviet Union and demand that ships carrying ballistic missiles to Cuba turn around and return home.

Afterward, Sherman Kent tried to justify himself by complaining that he was theoretically right but hadn't taken into account that Nikita Khrushchev could be so irrational; a more reasonable leader wouldn't have taken such a provocative action.

From my perspective, Sherman Kent was gripped by a flawed theory. His analysis of history left him complacent about Soviet intentions in Cuba. Kent not only had a bad idea, but he also held it tenaciously.

In contrast, John McCone was less trusting of Soviet intentions, which freed him to give the U-2 data a fresh look. He made judgments using the data, not his preconceptions. Or, rather, he had a preconception that was more skeptical and more accurate. Fortunately, McCone headed the CIA and could keep Kent's misjudgment from doing much damage.

The Israel Defense Forces provided a similar set of twins in the days leading up to the Yom Kippur War of 1973, another October crisis. Unfortunately for the Israelis, it was their chief of intelligence who was deluded, whereas his subordinates accurately picked up the warning signs of an attack.

Before October 1973, Israeli intelligence analysts were watching the Egyptians stage a military buildup along the Suez Canal under the pretense that it was merely a training exercise. Many in the Israeli intelligence community saw the threat. They worried that the Egyptians were lying and were not engaged in a training exercise, but they couldn't budge the top intelligence commander, Major General Eli Zeira, director of military intelligence. Zeira was fixated on the idea that the Egyptians wouldn't attack until they had air superiority. Because Israel still had air superiority, Zeira explained away all the contrary indicators.

We now know that even though the Egyptians didn't have air superiority, they planned to protect themselves from Israeli air strikes by violating an earlier armistice treaty and moving their anti-air weapons forward. Anwar Sadat, the Egyptian leader, wanted to regain control of the Suez Canal rather than defeat the Israeli military forces. Sadat's plan fell outside of Zeira's assumptions. As a result, the surprise worked. On October 6, 1973, the Egyptians sent 100,000 soldiers against the 450 Israelis spread out along the length of the Suez Canal.

Because Zeira was sure that the Egyptians wouldn't attack, he stifled efforts by his subordinates to warn the Israeli leaders. Even two hours before the war broke out, Zeira was trying to calm Israeli leaders, issuing an assessment that the probability of war was low.

Using hindsight, we can see that Zeira was gripped by a flawed theory, just as Sherman Kent had been. In the Israeli case, the evidence of trouble was even clearer than in the Cuban missile crisis. Several Israeli reserve generals viewed air photographs of the Egyptian and Syrian layouts on the morning of October 6. The generals immediately concluded that an Egyptian attack was imminent. The Egyptians launched it that same afternoon.

What makes this event even more disturbing is that before the attack the Israelis had an almost perfect picture of the Egyptian attack plans, military deployment, and intention to launch a war. True, the Egyptians prepared a deception plan—that their buildup along the Suez Canal was just part of a training exercise. However, Israeli intelligence analysts found no evidence of any real training going on and unsuccessfully tried to convince Zeira that the training was a ruse. More than a week before the war started, King Hussein of Jordan secretly flew to Tel Aviv. He warned Israeli Prime Minister Golda Meir of Syria's impending attack, probably in concert with an attack by Egypt, but his warning had no impact on Zeira.

The Israelis had a spy within the top Egyptian levels. The spy didn't warn the Israelis of the attack until a few hours before it happened. Even then, Zeira didn't believe the spy's report. Zeira relied on his doctrinal evaluation rather than on the signals flowing in.

In contrast, the junior Israeli intelligence analysts weren't gripped by Zeira's fixation. They saw that the Egyptians had put themselves in a position of relative strength that permitted them to attack. The junior intelligence analysts weren't trying to predict the strategy or the motivations of Anwar Sadat, and they could therefore be objective about the data showing an Egyptian military buildup.

These examples present a clear lesson: people gripped by a flawed theory can ignore, explain away, or distort evidence that could lead to insights. Therefore, we may be tempted to conclude that people should trust data, not their theories. We don't want people to fixate on their theories.

Except that we don't want people to fixate on data either. Several other twins failed because they trusted flawed data. For example, David Charlton rescued Corning's composite project by casting doubt on the data. The materials engineers who trusted the data raised a false alarm. Recall some of the other cases, not part of this contrasting twins study. Walter Reed was told to ignore the mosquito theory because it had been disproven, except that it hadn't. Barry Marshall's data showed no link between gastritis and *Helicobacter pylori,* at least for the first thirty patients, because the samples were discarded too quickly; fortunately, he maintained his belief in his theory despite the data.

There's no simple guidance here. Holding on to a flawed theory can be a mistake, but so can trusting flawed data. Tenaciously clinging to a belief despite contrary evidence can be a mistake, but so can prematurely discarding a belief at the first encounter with contrary evidence. All we can conclude is that we're likely to miss the insight if we rely on a flawed belief, either in a theory or in data, and we make it worse if we're pigheaded and fixate on that belief. As the saying goes, "It ain't what you don't know that gets you in trouble. It's what you know for sure that just ain't so."

The more central the belief is to our thinking, the harder it is to give up. These core beliefs anchor our understanding. We use them to make sense of events, to inquire, and to arrive at judgments about other ideas. And so we are much more likely to explain away any anomalies rather than revise our beliefs in the face of them. Clark Chinn and Bill Brewer cataloged some of the ways scientists try to explain away anomalies that contradict their beliefs.

- They question the reliability of the methods used to collect the data.
- They find a reason that the data aren't truly relevant.
- They seize on a possible contaminating factor—some different possible cause that might explain the finding.
- They make a small, often cosmetic change in their theory.

Anything to avoid making a shift in their theory. Paul Feltovich has found that physicians show the same tendencies, which he calls "knowledge shields," to hold on to their original diagnoses even when these diagnoses are wrong. The failure twins displayed many of these tendencies as they protected their cherished beliefs.

LACK OF EXPERIENCE

The second reason we might miss an insight is inexperience. Many of the failure twins just didn't have the experience to achieve the insight. Napoleon was an artillery officer and had been learning to use light artillery. He could recognize opportunities that his commanding officer missed. In the case of John Snow versus Chadwick and Farr, the proponents of miasma theory, Snow was a physician and had seen firsthand that cholera damaged the digestive system rather than the lungs. Chadwick was the London sanitation commissioner, and Farr was a demographer. No contest.

In reviewing the full sample of 120 cases, I judged that two-thirds of them depended on experience. People without the necessary experience couldn't have gained the insight. In the smaller set of contrasting twins, 17 out of the 30 failure twins simply lacked the experience to see the implications and gain the insight.

Experience isn't just about having the necessary knowledge. Experience is about how we use our knowledge to tune our attention. Our background can sensitize us to cues or patterns that others might miss. Think of Martin Chalfie hearing about the green fluorescent protein. His work with translucent worms gave him a different reaction than that of others hearing the same lecture. This notion of being attuned, of being sensitized, fits with the concept of a generally prepared mind. People with a generally prepared mind haven't done specific homework to get ready for their insight. Rather, their efforts and their interests have prepared them to notice things others miss.

John McCone had been worrying about the Soviet military buildup in Cuba even before the missile crisis. He had a hunch that the Russians might try something. His hunch didn't tell McCone what they would try, but it prepared him to look at the new data with a skeptical mind-set, just what you would want the head of an intelligence agency to do. McCone not only was free of Sherman Kent's flawed belief, but also had the right story for picking up on the weak signals that were there. The same is true for the junior intelligence analysts in the Israel Defense Forces. They became alarmed at the Egyptian buildup and the absence of Egyptian training events. Their Tilt! reflexes got triggered.

What about Sherman Kent and Eli Zeira? Surely they had lots of experience, at least as much as the others and, in Zeira's case, more experience than the junior officers had. Maybe flawed beliefs trump experience. Even when we have lots of experience, if we are gripped by a flawed belief, we will miss the insights sitting right in front of us.

A PASSIVE STANCE

The third reason we fail to gain insights is our stance. Many of the failure twins seemed passive. They took care of the necessary tasks but didn't actively scan for new developments and opportunities. Having an active versus a passive stance turned up in 21 of the 30 contrasting twins.* For example, the previous chapter described a case in which Devorah found the answer to the fantasy baseball question about what would happen when she passed her limit of 1,250 innings. I stopped thinking; Devorah, in contrast, started thinking even harder. Her mental motor kept running. Her attitude was active, open to unexpected possibilities.

In the set of contrasting twins, the twins with the active stance were often skeptical, ready to question the prevailing wisdom. The active-stance twins were open to data and inquiry, collecting new data. These twins often wondered what others were thinking and listened carefully to their explanations in order to see their perspectives. The twins with the passive stance didn't even inquire.

Consider the case of Ginger (she asked me not to use her real name), a very close friend of mine, who was frustrated by a legal problem that emerged when she changed jobs. She had signed a noncompete agreement when she had joined her previous employer right out of graduate school four years earlier. The agreement stated that if she left that company, she couldn't do any work or provide any services to the company's clients.

Ginger met with a legal officer at her new company, and he told her she was out of luck—she had signed the noncompete agreement, and she would have to live up to it.

* The totals come to more than 30 because some of the contrasting twins fit more than one of the four categories.

She was walking down the hall back to her desk feeling unhappy about the restriction but also baffled about how she was supposed to honor the agreement. How could she be sure to avoid all her previous firm's clients if she couldn't possibly know who they were? She remembered some of the ones she had worked with, but was sure she had forgotten others. And she certainly didn't know most of the firm's other clients during her time of employment. The firm had offices all around the world. How could she possibly be sure to avoid all their clients? This was so unfair. Tilt!

That's when she realized that her situation was unfair because she was trying to do something that was impossible. Without knowing all the clients, she had no way to avoid each of them. The only way to comply would be to have a complete list of clients, which led to her deeper insight about her previous employer: the company would never give her that list, especially as she was now working for a competitor. She had been thinking this was her dilemma, but it actually was her previous employer's problem! The noncompete stipulation was internally inconsistent. She could use the inconsistency to free herself from the noncompete clause.

In a much better mood, Ginger called the compliance officer at her previous company. She assured him that she intended to honor her agreement, and he was happy to hear that until she explained the difficulty she was having because she didn't know all the company's clients during her period of employment. She asked if he would send her a complete list of every client for the past four years. Citing confidentiality concerns, he naturally refused. Ginger explained that without such a list, she couldn't live up to the agreement. There was silence on the line as the compliance officer thought about it. Finally he said, "Then don't worry about it."

Ginger found a way out of her dilemma, but the legal officer at her new company missed it. He understood the ramifications of the agreement she'd signed, and that's as far as his thinking went. He viewed his job as interpreting legal clauses. Ginger, on the other hand, kept mulling over the implications and the constraints and found an escape hatch.

An active attitude leads to persistence. The successful twins were more persistent. They seemed to tolerate failure better and weren't necessarily deterred by showstoppers. Barry Marshall persisted in his study of ulcers even though the first thirty patients didn't show any signs of *Helicobacter pylori* infection.

A CONCRETE REASONING STYLE

People differ in how well they tolerate contradictions and ambiguity, and this personality style likely affects their success at gaining insights. People also differ in how ready they are to entertain ideas that they don't think are true and in how much they enjoy imagining alternative universes.

Some people become impatient with speculation. They see the playful exploration of ideas as a sign of immaturity. They want closure, and they roll their eyes when a member of the group starts going off on tangents. They are concrete thinkers who just want to work with the facts, not with flights of fancy. This concrete reasoning style wouldn't leave people very open to insights.

The playful reasoning style likes to juggle ideas and imagine hypothetical scenarios. Meredith Whitney wondered what it would look like if Bear Stearns were in financial straits, and she tailored her investigation with that premise even though she doubted it at the start. This personality trait is about having a playful mind and enjoying speculations about hypothetical situations. It is different from having an active or passive stance, which can vary from one situation to another. The playful versus concrete reasoning style is a relatively fixed personality trait.

I judged that in almost half my sample of contrasting twins, 14 out of 30, the failure twin was more prone to concrete thinking than the twin who had the insight. This was a tough dimension to rate, but some cases stood out clearly: Napoleon versus General Carteaux, John Snow versus the miasma spokesmen Chadwick and Farr.

Flawed beliefs, limited experience, a passive attitude, and a concrete reasoning style. A quadruple whammy that stifles insights. Yet we have to be careful here. While these four reasons may get in the way of achieving an insight, engaging in the positive behaviors won't guarantee our success. The cases I've drawn on were all success stories. There are, of course, many instances in which people rooted out all shaky beliefs;

worked hard to develop expertise; had an active, persisting attitude; speculated like crazy; and still got nowhere.

What isn't included in the four factors that stifle insights? Is intelligence a factor? I don't think so, although my daughter Devorah might disagree. I wasn't able to flag any differences in intelligence because most of the 30 cases required a reasonable amount of intelligence just to get in the mix as a contrasting twin.

THE DOUBLE HELIX: TRIUMPHS AND FAILURES

Most of the contrasting twins, 24 out of 30, differed on more than one of the factors of flawed beliefs, lack of experience, passive attitude, and concrete thinking style. Consider the way James Watson and Francis Crick discovered the molecular structure of the gene in 1953, one of the most important scientific discoveries of the twentieth century. Their double helix model explained the genetic secret of life and was the foundation for modern biology. Watson and Crick made insights into the nature of the genetic code that other scientists with the same data missed.

Watson and Crick bumped into each other at Cambridge University in the fall of 1951, two very junior researchers at one of the most prestigious schools in the world, both working at the famous Cavendish Laboratory. James Watson, only twenty-three, had already gotten his PhD. His interest in the gene started when he was a senior at the University of Chicago and continued through graduate school at Indiana University. After completing his PhD, he wound up at Cambridge to do postgraduate research on the structure of the protein hemoglobin. Francis Crick was even more junior than Watson. At the age of thirty-five Crick was still a graduate student, a physicist retraining in biology. With hindsight, they could be thought of as the Woodward and Bernstein of biology.

In casual conversations, Watson and Crick found that they shared the belief that genetic information was carried by DNA. They decided to work together on the riddle of DNA as a side project to their official tasks. What started as part-time work soon captured all their attention, and their official tasks faded into the background. They figured it was only a matter of time before someone solved the riddle. They wanted to get there first.

DNA (deoxyribonucleic acid) consists of four bases: guanine, cytosine, adenine, and thymine. We now know that these four "letters" make up the genetic code for all life. The sequence of these bases carries the instructions for the sequence of amino acids in proteins. But how does DNA faithfully copy the code from one generation to the next? Watson and Crick hoped that by understanding the structure of DNA, they might have clues to how it worked.

The two men had a hunch that gave them an advantage over most other researchers. Shortly before coming to Cambridge, Watson had attended a scientific meeting in Naples and heard a lecture describing DNA as having a repeating structure, which suggested to him that the structure of DNA could be decoded. Just as Martin Chalfie was stimulated during a lunchtime seminar by an unexpected set of comments about a green fluorescent protein in jellyfish, so was Watson excited by the chance to pin down the structure of DNA. Later, in Cambridge, Watson shared these ideas with Crick. They knew that the brilliant chemist Linus Pauling (who would go on to win two Nobel Prizes) had recently used X-ray diffraction methods to demonstrate that protein molecules were structured as a single strand helix. Watson and Crick speculated that DNA might also be some sort of helix.

They felt that a single helix wouldn't be complex enough. DNA would likely be a double or a triple helix. They decided to construct physical, three-dimensional models of each possibility. The models they built had to conform to the features of DNA that were known through X-ray crystallography photographs.

After much effort, and some embarrassing failures, they had gotten nowhere and had been warned by Sir Lawrence Bragg, the head of the Cavendish Laboratory at Cambridge, to give up their efforts. But then Watson got to see a photograph, taken by another researcher, showing clearly that DNA had a helical structure. The other researcher was skeptical about the idea of a helix. Because Linus Pauling himself was starting to examine DNA, there was no time to lose. Watson and Crick decided to make one last effort to model the structure of DNA.

Crick worked on building a model of the triple helix form of DNA and suggested that Watson work on the double helix form. However, Watson didn't have the three-dimensional building blocks he needed. While he waited for the shop to make these, he played around with some two-dimensional cutouts. He was startled to find that an adenine-thymine pair held together by two hydrogen bonds had the identical shape as a guanine-cytosine pair held together by at least two hydrogen bonds.

In that instant the double helix model sprang to life. Watson had solved eight mysteries simultaneously. He knew the structure of DNA: a helix. He knew how many strands: two. It was a double helix. He knew what carried the information: the nucleic acids in the gene, not the protein. He knew what maintained the attraction: hydrogen bonds. He knew the arrangement: The sugar-phosphate backbone was on the outside, and the nucleic acids were on the inside. He knew how the insides matched: through the base pairs. He knew the arrangement: the two identical chains ran in opposite directions, so they were mirror images. And he knew how genes replicated: through a zipperlike process. The diagram illustrates the double helix model of DNA.

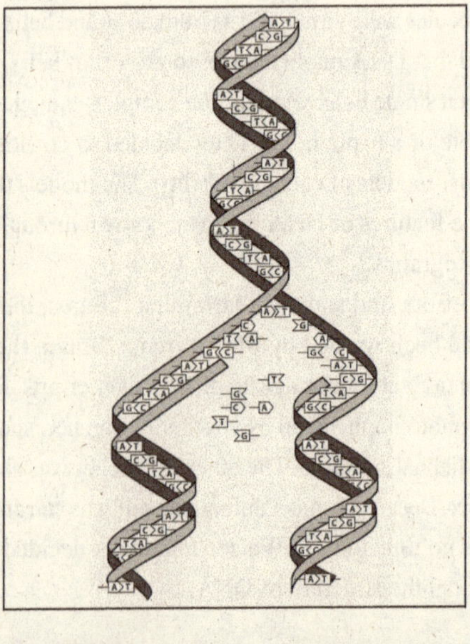

The discovery of the structure of DNA was only one of the thirty contrasting twins in my research, but it actually contains four twins within the same story, and each shows some of the four factors (flawed beliefs, limited experience, a passive attitude, and a concrete reasoning style) at work.

Somehow, Watson and Crick, two junior scientists, scooped the rest of the research community. Somehow, they saw implications in the foundational data on DNA collected by Murray Chargaff, a biochemist at Columbia University, that Chargaff himself missed. Somehow, they saw implications in Rosalind Franklin's X-ray crystallography data on DNA that she missed. And somehow, they succeeded, even though Watson never fully grasped what Crick was thinking until he pushed those two-dimensional cutouts together. The story of Watson and Crick isn't just about the insights they gained. It is also about the failures of others, the contrasting twins, to gain those insights.

The Scientific Community Versus Watson and Crick

The scientific community was hampered by the first factor we've covered in this chapter, flawed beliefs. Before the discovery of the double helix, most researchers believed that genetic information was carried by proteins because proteins have complex structures. In contrast, DNA, which consists of four bases (adenine, thymine, guanine, cytosine), seemed too simple to carry the genetic code. DNA was thought to play a secondary role, perhaps as a backbone for the real process. Watson and Crick, both new to the field, hadn't spent years in the grip of the flawed belief that protein, not DNA, was likely to carry the genetic code.

The bacteriologist Oswald Avery (1944) had reported a study showing that bacterial DNA embodied the bacterial genes, but most researchers thought Avery's work lacked the necessary controls, and even Avery was reluctant to believe his findings. The leading researchers at the time, Max Perutz, Max Delbrück, and Salvador Luria, all dismissed the implications of the Avery data.

Oddly, the scientific community was also hampered by the second factor, inexperience. Watson and Crick were the upstarts. They should have suffered for their lack of experience. However, many of the leading geneticists at the time didn't know biochemistry or care about it; they were just interested in the characteristics of genes. The organic chemists who were studying DNA weren't interested in genetics.

In contrast, Watson and Crick were a good collaborative team. They had the right blend of experience and complemented each other. Crick's background was in physics, X-ray diffraction methods, proteins, and gene function. Watson's background was in biology, phages (bacterial viruses), and bacterial genetics. Crick was one of the few crystallographers interested in genes. Watson was the only one coming out of the U.S.-based phage group who was interested in DNA.

I don't believe that youth per se gave Watson and Crick an advantage. Rather, they weren't gripped by the flawed belief that proteins carry the genetic information. They came along just at the right time. In 1952, a year before Watson and Crick published their article on DNA, Alfred Hershey showed that when a phage particle infects its bacterial host cell, only the DNA of the phage actually enters the cell. The protein of the phage remains outside. This finding contradicted the belief that genes are proteins. It also suggested a connection: that DNA may play a role in genetics. The researchers who were studying proteins to learn about genetics had to shift gears. Watson and Crick started out in the right gear.

Chargaff Versus Watson and Crick

Erwin Chargaff had recently found that in DNA the amount of adenine matches the proportion of thymine and the amount of guanine matches the amount of cytosine. Watson and Crick weren't fully aware of this work. Fortunately, in July 1952 they had a chance to go to lunch with Chargaff, who described his results to them. Crick was electrified. He immediately saw the implication that DNA used complementary pairings. Thus, a molecule of adenine could be copied using a molecule of thymine (complementary pairing) rather than another molecule of adenine (like-with-like pairing).

Chargaff came away from the lunch with an unflattering impression. He considered Watson and Crick dabblers. They weren't collecting data, and they didn't have much of a background—Crick was still a graduate student. Besides, they were supposed to be working on other projects. They babbled about a helix but didn't know much of the literature, including Chargaff's own papers. Crick couldn't even remember the names of the four bases.

Why didn't Chargaff see the implications of his own research? To the end of his career, he remained bitter about this lost opportunity. He wasn't gripped by a flawed belief, and he had much more experience than Watson and Crick. That leaves the third factor, passivity. Chargaff was content with his research program. He didn't seem to have the active attitude of Watson and Crick to inquire about what his own discovery might mean.

In contrast, Crick, in particular, had an active mind-set. He wasn't just trying to learn how DNA was shaped. He also wanted to know how it works. Almost from the beginning he was imagining how DNA could code and copy genetic information. So when he met with Chargaff, he immediately saw an implication of the ratios that Chargaff had found. The ratios meant that the bases could complement each other. The strands of DNA could copy themselves using this complementary feature. They didn't have to use a like-with-like mapping. Chargaff didn't take this active, functional perspective and missed the implication of his own results.

Franklin Versus Watson and Crick

The contrasting twin who was most similar to Watson and Crick was Rosalind Franklin, one of the leading X-ray crystallographers of the day. She was actually gathering data on DNA in London, an easy train ride from Cambridge, where Watson and Crick were working. Her mission was to unravel the structure of DNA. Watson and Crick were poaching on her territory. She is the scientist who collected the X-ray diffraction data that galvanized Watson and Crick. Her supervisor, Maurice Wilkins, had surreptitiously shown this photograph to Watson, sending Watson and Crick off on the final leg of their journey.

Franklin was close, but she was still a few steps away. Many have speculated about why Franklin didn't take those final two steps to discover the structure of DNA. I have nothing new to add to this controversy. I merely want to view Franklin's failure using the four factors I have introduced in this chapter.

First, Franklin was gripped by a false belief: that DNA did *not* have a helical structure. Her flawed theory stemmed in part from flawed data reported by an earlier researcher who mixed up the wet and dry forms of DNA. Franklin fixated on her flawed idea; it became her quest to show that DNA was not a helix. Even late in the game Franklin thought she had disproven the helix notion and titled a short note "Death of DNA Helix."

Consider Franklin's famous photo number 51 that Wilkins showed to Watson. Franklin let that photo sit for ten months before she saw that it revealed DNA to have a helical structure. Watson immediately understood the implication. "The instant I saw the picture my mouth fell open and my pulse began to race. . . . The black cross of reflections which dominated the picture could arise only from a helical structure." He was already thinking about a helix. Photo 51 confirmed his hunch, plus the photo showed the spacing and dimensions that he and Crick needed for their model. Watson was tuned to look for evidence of a helix; Franklin was tuned to look for evidence against a helix.

Also, in an annual report on her research, Franklin presented several data points about her DNA findings. When he read Franklin's report, Crick immediately saw that these findings implied that the chains run in opposite directions. Franklin, who wasn't thinking about a complementary pairing, didn't see the implication. (Neither did Watson or anyone else at the time.)

Second, Franklin lacked certain types of experience. She didn't come out of a model-building tradition the way Watson did. Therefore, she wasn't thinking about the distances and angles in the DNA molecule, whereas Watson and Crick needed to understand the details about distances and angles in order to build a model of the molecule.

Third, Franklin took a passive view of science. She disliked the eagerness with which Watson and Crick speculated about questions that, she thought, should best be resolved through patient data gathering. Her work as a crystallographer gave her a more static perspective than that of Crick, who couldn't stop himself from speculating about how DNA replicated.

The fourth factor also was operating here. Watson and Crick were each very playful reasoners. They relished the model-building exercise that allowed them to try out different ideas. The other scientists in this story were high in hypothetical reasoning, but it seems unlikely that they reached the level of Watson and Crick, who were so willing—eager—to speculate about hypothetical situations. Other researchers saw such rampant speculation as immature and preferred to carefully collect data to add to the scientific edifice.

Watson Versus Crick

During the lunch with Erwin Chargaff, Crick got excited when he learned of Chargaff's findings. But Watson wasn't paying attention. He didn't grasp the connection Crick made. Even afterward Watson couldn't warm up to the notion of complementary pairing. His mind was fixated on a like-with-like pairing mechanism. Thus, Watson was gripped by a false belief. His thinking was also passive on this one issue. He was sufficiently content with the idea of like-with-like pairing that he didn't expend any mental energy looking further.

A reading of Crick's autobiography and other accounts, not just Watson's famous book, *The Double Helix,* shows that Crick had a somewhat different hunch than did Watson. Crick was using a functional frame—how DNA might work—and was trying to imagine how the DNA structures they were designing could replicate genetic information. Watson appeared to have a more structural frame. He wanted to capture the structure of DNA and wasn't thinking so much about how that structure would code for information. He barely remembered the meeting with Chargaff until Crick later reminded him.

Therefore, on at least this issue Watson and Crick were matched contrasting twins who had exactly the same information but differed in what implications they drew. According to Crick's account, the key to the discovery of the genetic code was the base pairings and not the helix shape for DNA. Watson's account focused on the helix. He titled his book *The Double Helix*. He didn't share Crick's enthusiasm for complementary pairing until he stumbled upon it himself.

And let's not forget about luck. I haven't acknowledged luck as a fifth factor that distinguishes winners from losers, but luck certainly played a role in Watson and Crick's discovery of DNA's role in genetics.

The four DNA bases are flat, so Watson really could model them in two dimensions. He was doodling with the 2D representations, waiting for the machine shop to finish the 3D forms, when he stumbled upon the complementary pairings. If the molecular structure of the bases weren't flat, Watson's doodling might not have worked. And if he had waited for the parts for the three-dimensional model, Watson might not have noticed the complementary pairing.

Rosalind Franklin had some bad luck. DNA can take two forms, dry and wet. Rosalind Franklin relied on another scientist's data that mixed these forms, which threw her off. Then she concentrated her research on the dry form. If she had instead chosen to investigate the wet form, she would have had a better chance. And then Franklin spent some time studying an unusual DNA variant that turned out to be a false trail.

All the researchers struggled with bad data; Watson and Crick were lucky to recover more easily than the others. For instance, Watson and Crick were using one configuration for bases, but they should have been using a different configuration. They happened to encounter a colleague who explained to them that the textbook data for hydrogen bonds were wrong! All at once, Crick's aversion to using hydrogen bonds in their model disappeared. The correction of the base configuration also confirmed the idea of complementary pairing to Crick; the new configuration was incompatible with like-with-like base pairing.

We have trouble overcoming the four factors—flawed beliefs, insufficient experience, a passive stance, and a concrete reasoning style—that emerged from the contrasting twins study. One of them, the playful/concrete reasoning style, is baked into our minds as part of our personalities.

We also at times make things even more difficult by erecting additional barriers, such as the computer systems we use. Software designers don't deliberately try to stifle our insights, but some of the methods they use have unintended consequences.

CHAPTER ELEVEN

Dumb by Design

As computers have proliferated throughout our lives, information technology professionals have come up with guidelines for how to make these versatile machines more helpful. Software developers, information management specialists, human factors experts, and many other communities have developed recommendations for designing decision support systems and information management systems that are easy to use and that enable us to be more successful.

I have selected four of the more common guidelines. They impress me as useful for designing computer-based aids. The question we'll explore in this chapter is, What effects do these specific guidelines have on insights? Here they are:

1. The system should help people do their jobs better.
2. It should clearly display critical cues—the items of information that users rely on to do their jobs.
3. Conversely, the system should filter out irrelevant data so that the operators aren't drowned in meaningless messages.
4. And the system should help people monitor progress toward their goals.

These guidelines seem pretty reasonable and compelling. But let's see how useful they would have been in a situation that called for insight: the way Daniel Boone foiled the kidnapping of his daughter and two of her friends. As we explore this example, think about how we could use the four design guidelines to help Boone do his work.

RESCUING JEMIMA

Daniel Boone was a hunter. He is famous for many things: exploring and settling Kentucky, fighting off attacks by Shawnee Indians, blazing the Wilderness Road from North Carolina and Tennessee into Kentucky so that several hundred thousand settlers could migrate to western Virginia and Kentucky, and founding Boonesborough, Kentucky. So it is easy to forget that for most of his life he made his living by hunting. Born in 1734, Boone got his first hunting rifle when he was twelve, and set out on his last hunt in 1817, three years before he died. Most of his hunts were over in a few days, but others—the long hunts—lasted up to two years.

Hunting isn't just about shooting straight. Hunters have to track a quarry. If the quarry can move faster than the hunter, however, tracking isn't enough. The hunter will need to anticipate where the quarry is heading in order to get there first. And anticipation often depends on insight.

The skills of tracking and anticipating came into play in 1776 when Boone had to find the raiding party that carried off his daughter. By the late spring of that year the European American population living in Kentucky had dwindled to about 200 because so many settlers had fled after a series of Indian attacks. One of the remaining settlements was Boonesborough, established on a bank of the Kentucky River in 1775.

In early July 1776, a small war party of two Cherokees and three Shawnees infiltrated the region around Boonesborough and spent more than a week exploring it. The Indians murdered a farmer who lived several miles away, but mostly they scrutinized Boonesborough, looking for an opportunity.

On Sunday afternoon, July 14, 1776, they found it. Jemima Boone and two of her friends took a canoe for an innocent outing. Jemima was thirteen, the fourth of Boone's ten children; the other two girls were fourteen and sixteen. Jemima had injured her foot and wanted to bathe it in the Kentucky River. Daniel Boone warned his daughter to stay close to the shore, but her friends teased her. Betsy Calloway, Jemima's sixteen-year-old friend, impetuously steered the canoe out into the center of the river, where it was carried away by the current. The raiding party saw its chance. The Indians rushed downstream.

When the canoe came close to the far shore, one of the warriors jumped into the river and caught hold of a strap attached to its bow. Fourteen-year-old Fanny Calloway, Betsy's sister, started pounding the warrior's head with her paddle, and the girls started screaming. The raiders quickly pulled the canoe to shore, dragged all three girls into the forest, and threatened them with a knife if they didn't quiet down.

The screams of the girls alerted the settlers. Daniel Boone was enjoying his Sunday afternoon nap. On hearing the tumult, he jumped up, pulled on his pants, and rushed out barefoot. He hastily formed a rescue party to pursue the kidnappers, who had left the girls' canoe behind and were fleeing on land.

The first order of business was getting across the Kentucky River. Richard Calloway, father of the two kidnapped sisters, led a group of horsemen a mile downstream to a shallow point where they could cross the river. Someone else swam across the river to the abandoned canoe, paddled back, and ferried Boone and the others to the opposite shore.

Now Boone had to pick up the track of the kidnappers. He sent some of his group upstream while he took the rest downstream. Boone didn't find any signs, but he did run into Calloway and his horsemen. They were impatient to search for the Indians. With their horses they could move faster than the Indians, and Richard Calloway was desperate to find his daughters.

Insight #1: Redirecting the horsemen. Horses can move faster than men in open country, but the Indians had fled into the forest, which reduced the riders' advantage. Boone imagined how they were going to conduct a search as the daylight was fading. They had no idea which way to go. It seemed futile.

Then he had a different idea. He told the riders not to go searching.

Boone knew that most of the conflict in Kentucky had been with the Shawnees, so it was likely that the raiders were Shawnees. There were no Shawnee villages nearby; the major Shawnee villages were in Ohio, north of the Ohio River. Boone himself had been captured by Shawnees in 1769. His captors tried to bring him to Ohio, but Boone escaped as the Shawnee tried to cross the Ohio River. So Boone imagined that these kidnappers were heading north to the Shawnee towns. He assumed that when the kidnappers started out, they had paddled across the Ohio River and stashed their canoe on the Kentucky shore, proceeding to Boonesborough on foot. They'd probably return the same way. That meant the Indians had to get across the Licking River, south of the Ohio River. Without a canoe, the best place to cross the Licking River would be the shallow stretch at the Upper Blue Licks.

This was Boone's insight: he anticipated the path the Shawnees were going to take, and he imagined a specific location where Calloway and his riders could ambush them. Boone's insight included a plan for the riders not to search for the Shawnees but to gamble on surprising them as they crossed the Licking River. He used the creative desperation path in the Triple Path Model to abandon a weak assumption: that the riders had to search for the raiding party. They didn't need to search—they needed to ambush.

Boone's insight gave Richard Calloway a new story. Calloway now understood why the Indians were going to the Upper Blue Licks. He acquired a new action—the capability of ambushing them. His attention shifted to navigating to the ambush area and then finding places to hide while he and his riders waited. His feelings of confusion and desperation were replaced by determination. And his goal transformed from searching to ambushing.

In the meantime, the upstream group of rescuers had located the kidnappers' trail and started to follow it. After sending Calloway off, Boone and the others joined the search, relying on Boone's tracking skill.

The girls contrived to leave fragments of their clothing and ribbons to mark their trail. By the second morning of the chase, Boone and his followers were about ten miles behind the kidnappers, but the trail was getting more difficult to follow.

Boone judged that the Shawnees' lead was growing. So it was unlikely that he and his party would be able to catch up to the kidnappers. Boone felt discouraged.

Insight #2: Abandoning the chase. The Indian raiders had been moving north, just as Boone expected. He was pretty confident that they were planning to bring the three girls across the Ohio River and into Ohio, and that they were currently heading to the Upper Blue Licks. Therefore, Boone switched tactics. Instead of persisting in a futile chase, Boone decided to lead his party directly to the Upper Blue Licks, where Richard Calloway and his group of riders were waiting. The men were very uncomfortable abandoning the trail, but they followed Boone's orders. Boone probably thought his band of rescuers might catch up to the Indians before they reached the Upper Blue Licks.

Boone's second insight was another case of creative desperation, jettisoning the assumption that they would be able to catch up with the Indians and understanding how they would be able to intersect the war party. Insight #2 shifted the story in many of the same ways that insight #1 had. Boone understood that the Indians were moving too fast to be tracked but they were also conforming to his predictions, so he could switch from tracker to anticipator. His attention focused on reading the terrain to get to the Upper Blue Licks as quickly as possible; he noted signs of the raiding party but resisted the temptation to resume tracking. His spirits had sunk when the tracking didn't go well but lifted when he adopted the new strategy. And his goal transformed from tracking to intersecting the Indians.

Boone's group could now move faster than the Indians because they didn't have captives and they had a greater sense of urgency than the kidnappers. By the third day, Boone and his party were spotting signs of the kidnappers, suggesting that they were only an hour ahead: muddied water as they crossed a stream, a dead snake, and the carcass of a freshly butchered buffalo calf whose hump still oozed blood.

Insight #3: Cooking the meal. The fresh carcass of the buffalo calf suggested to Boone that the Indians were getting ready to cook the meat. Further, he expected them to do the cooking when they came to the next source of water. That was his insight. When the rescue party came to Bald Eagle Creek, Boone recognized it as the water source the Indians would want to use.

Boone again divided his party, half going upstream and the other half downstream. The upstreamers located the raiding party. Boone's third insight altered his understanding; the Indians were nearby and were going to be vulnerable as they made camp. Now he could take a different action—to surround and attack the kidnappers. He was looking differently, searching for a source of water and then for the Indian camp. Boone's emotion changed from determination to excitement. His new goal was to surprise the raiding party.

The Indians had made camp and were relaxing, roasting the buffalo meat over a fire. Boone had been right, using the buffalo carcass to form a connection insight about what the Indians were going to do next.

The war party had posted a rear scout to make sure no one was following them. If Boone had been tracking the Indians, he probably would have been spotted. But he wasn't following them. He was relying on anticipation to trap them.

Boone's group surrounded the camp and prepared a surprise attack. Jemima spotted her father creeping forward on his belly—he signaled to her to be quiet. Then the settlers opened fire. Two of the girls had the sense to drop to the ground and lie flat, but Betsy, ever the scatterbrain, jumped up at exactly the wrong moment. Suddenly that rude description of Betsy threatened to take on a literal meaning. One of the Indians tried to club her skull in but missed—years later she described to her children how she had felt the war club graze her head. The Indians fled without their weapons; two eventually died from their wounds. The settlers didn't bother giving chase. They were content to have rescued the three girls unharmed except for scratched legs and shredded clothes.

Now let's consider how we could have helped Daniel Boone by giving him some modern technology. How would we design that system?

Guideline 1: The system should help Boone do his job better.

Such a system would have registered Calloway's job as catching up to the Indians and Boone's job as tracking the Indians, and then worked to help them perform their jobs effectively. And all this support would have become useless once Calloway and Boone switched tactics. It would have been worse than useless, requesting types of information that would have been irrelevant. Boone would have turned the system off once he stopped trying to track the raiding party. He would have tossed the system into the bushes in order to move more quickly.

The goal of helping people perform their jobs better makes sense only if they have well-defined and stable jobs. It doesn't apply to people whose jobs may shift as a result of insights. The job of the riders shifted from searching to ambush. The job of the settlers in Boone's party shifted from tracking the Indians to intersecting them. The system we design to help people do their current jobs may lock them into the original job concept and reduce the chance for insights. This guideline can make it harder for them to rethink their jobs. The system would have penalized Calloway and Boone for changing tactics. They would have had to take precious time to do some reprogramming. People sometimes find it easier to stick with the original job description to avoid the hassle of making changes.

Guideline 2: The system should clearly display critical cues.

This guideline isn't much better. When we identify critical cues in advance, we'll likely miss those that unexpectedly become relevant after we have an insight. The cues the riders needed for searching had nothing to do with their new ambush mission. The cues Daniel Boone needed to track the Shawnees became irrelevant once he decided to stop tracking them. Many computer aids rely on getting information into and out of databases, and these databases are organized so that the users can navigate them without getting lost. However, the original database structure is likely to become obsolete as knowledge workers gain more insights and revise their thinking. When that happens, the original database structure becomes cumbersome and may get in the way of insights.

Guideline 3: The system should filter out irrelevant data.

This guideline is actually harmful. The riders needed navigational help to reach the Upper Blue Licks, but the system would have screened these cues out as having no value for their search. The cue of the buffalo carcass would have been downplayed as part of a set of indicators that they were closing in on the raiders, rather than highlighted as a central anchor in a new insight. A system that tried to help Boone might well have obscured the newly critical cue.

The recommendation to filter out irrelevant data sounds good as long as we know in advance which data are relevant. It screens out the happy accidents and unexpected cues that often spark insights. Think about some of the other examples we've covered. Prior to the Battle of Taranto, the naval attack the British launched against the Italian fleet early in World War II, neither Admiral Stark nor Admiral Yamamoto, the American and Japanese naval commanders, would have defined "air attacks from aircraft carriers" as a relevant data field in assessing the safety of a fleet of ships anchored in a shallow bay. Jocelyn Bell Burnell, who unexpectedly discovered pulsars, wouldn't have configured her digital recording system to catch the unusual squiggles; the automated data collection would never have even alerted her to those squiggles.

The design recommendations wouldn't have helped Napoleon dislodge the British at Toulon. If Jean François Carteaux, the commander of the French forces, had directed his staff to prepare a French Republic Army Decision Aid, it would have been designed to overpower the British and Spanish forces. It would have displayed the critical cues for capturing Toulon, such as weak points that could be attacked, relative strength of the French versus the Anglo-Spanish armies, and optimal placement for the French heavy artillery. The Carteaux decision aid wouldn't have helped Napoleon; it would have gotten in his way. It would have filtered out "irrelevant" data such as the location of small forts, like l'Eguillette and Balaguier, that were too far away to threaten the Anglo-Spanish army.

The desire to filter out irrelevant data, guideline 3, is understandable. Most of us feel we are drowning in data, so we're grateful for any help we can get. Unfortunately, this guideline is responsible for creating Internet filter bubbles.

A powerful search engine such as Google can be too helpful. It learns our preferences and shows us matches that are likely to satisfy us, but we don't get exposed to other kinds of options. In political arenas Google finds articles that match our beliefs, screening out opinion pieces from other parts of the spectrum. And we never see what is filtered out. Google is wrapping us in a cocoon of our own beliefs.

Eli Pariser, a left-wing political activist and the board president of moveon.org, first noticed Internet filter bubbles when Facebook deleted all his links to conservative friends. He had neither been asked for permission nor even been told about it, discovering the deletions by accident. Facebook had picked up his trend of following liberal friends more than conservative ones and decided to help him out. However, Pariser wanted to stay in contact with conservatives. He hadn't noticed how much he was favoring his liberal friends. But Facebook did.

Systems such as Google determine what we don't want to see and either filter it out completely or bury it so deep, perhaps on page 25 of the search results, that we probably won't find it. The personalized searches we get from Google, Yahoo, and others gauge our preferences and then screen out the items we're likely to find irrelevant. Pariser argues that searches also need to show us items that are challenging and even uncomfortable. They need to expose us to other points of view.

Guideline 4: The system should help people monitor progress toward their goals.

This guideline also creates problems. Such monitoring will help us stay on schedule. But progress monitoring can get in our way if, after we start a project, we have insights about how to restructure the tasks. Even worse, what happens if we have insights about how we should revise our goals? Once we give up the original goals, all our progress markers become obsolete. A tool that monitors progress toward the original goals would have pestered the riders for conducting an inefficient search strategy. It would have hounded Daniel Boone for not following the optimal track. It would have warned him to hurry up when he stopped tracking the Shawnees. It would have downgraded his performance for dallying at Bald Eagle Creek instead of moving rapidly to the Upper Blue Licks. If we want to discourage insights, we can't do much better than this feature.

Insights change our *understanding* by shifting the central beliefs—the anchors—in the story we use to make sense of events. The Triple Path Model shows that our new understanding can give us new ideas about the kinds of *actions* we can take; it can redirect our attention, changing what we *see;* it can alter the emotions we *feel;* and it can affect what we *desire.* This last outcome is the one most disrupted by guideline 4.

A system that keeps us on track to reach our original goals is likely to interfere when we discover new goals. Such a system would have harassed Jocelyn Bell Burnell for wasting time on her squiggles, time that should have been spent searching for quasars. It would have complained when Walter Reed and his team started chasing after mosquitoes instead of improving sanitation and increasing air quality. It would have given Wagner Dodge feedback about the running speed he needed to maintain and might have sounded an alarm when he stopped running to light his escape fire.

STRONGER DESIGN = WEAKER INSIGHTS

Each of the four design principles listed at the beginning of this chapter depends on order and structure, whereas insights are disorderly. To change the way decision aids and information technologies are designed, developers would have to pay less attention to how the work was done in the past and more attention to giving decision makers room to discover—giving them more freedom to modify their tasks. System designers should make it easy for users to shift goals and plans without getting disoriented.

I don't imagine that advice would go over well. Software developers might recoil from such a suggestion and view it as a call for anarchy. Therefore, I am not optimistic about the chances for designing systems that foster insight. The four guidelines at the beginning of this chapter are too compelling. In the past, when I led seminars on cognitive systems engineering, I advocated for these guidelines. I was a believer until I started my project to investigate insight. Don't get me wrong. I still consider the four guidelines useful. However, I now see how they can interfere with insight by freezing users into obsolete job practices.

Even if system designers wanted to create more flexibility, their sponsors and their organizations might resist. In designing an information technology application, a developer must have clear goals, objectives, and benchmarks.

If rigidly designed information technologies can sometimes get in the way of insights, that's nothing compared to what organizations do. Organizations insist that they value innovation, but you wouldn't know it by studying their actions.

CHAPTER TWELVE

How Organizations Obstruct Insights

ORGANIZATIONS INADVERTENTLY SUPPRESS the insights of their workers, and they do so in ways that are ingrained and invisible. This chapter will first explore the motivations—the reasons that organizations engage in this suppression. Then we'll examine the methods—the ways that organizations block insights.

THE MOTIVATIONS

Organizations stifle insights because of forces locked deep inside their DNA: they value predictability, they recoil from surprises, and they crave perfection, the absence of errors. Surprises and errors can play havoc with plans and with smooth operations. In their zeal to reduce uncertainty and minimize errors, organizations fall into the predictability trap and the perfection trap.

The Predictability Trap

Imagine that you are given a fairly important and complex project to manage. You want to get the best work from your project team. You want team members to gain insights that will make the project a big success. You tell them so. You believe it yourself. You've read the books on innovation and brainstorming. You've guzzled the Kool-Aid of creativity. You can't wait to put all those inspiring ideas into practice.

Yet you also want to make sure the project is completed on time and within budget. You can't tolerate a plan that contains one of those "and then a miracle occurs" steps in the middle. Therefore, you carefully map out the steps that will carry you successfully from start to finish. You assure yourself that you are not locking the project into a plan; you have to set up a plan to give yourself some guidelines, some platform from which to innovate. Besides, your project won't get approved unless you can enter a plan into the computer-based management system that tracks each new effort.

You set up a timeline showing when each step starts and finishes. You calculate the resources you plan to allocate to each step—the dollars you will spend and/or the hours you will need. You name the people responsible for each step. Now you relax. Your plan should carry you and the team to success. Even when the plan gets disturbed, you'll be able to quickly detect the perturbation and reprogram resources so that progress isn't slowed. Notice that as the program manager, you are spending most of your time monitoring and tweaking the plan. You aren't thinking about insights.

A team member walks into your office and says, "You know, I've been wondering, what if we shift some people around? Our group really needs some of the skills from the folks in another group. We're always on the phone with them." "Interesting idea," you tell him. "I'll give it some thought." But your thoughts are mostly on the three phone calls you have to return and the briefing you have to prepare. Besides, reassigning workers is not a trivial issue. Their supervisors get angry, and the workers themselves worry about why they've been thrust into a different unit. You feel annoyed by this intrusion, but you calm down by telling yourself that your visitor, despite his good intentions, doesn't have the experience to anticipate the problems his suggestion could cause.

Another team member has an even better idea. She has a real insight about how to modify your project plan, combine several tasks, and produce something your customer might find more useful. This time you act on it. You take her suggestion to your director. The expression on his face flits between pity and contempt. He patiently reminds you—in that slow-talking style that makes it seem as if he were dealing with someone who is mentally handicapped—that the contract was written to conform to the project tasks. Progress payments are linked to the timetable for completing each phase. If you shift the schedule, the automatic payments will get suspended until you can get a new plan approved, a process that might take months.

You have learned that insights about tasks and goals threaten your relationship with your director. They weaken your credibility. You walk out of his office berating yourself for being so foolish and vowing not to make that mistake again, not to rush to him with any more bright ideas.

Your director has started to look at you as someone he'll have to keep an eye on, no longer a person who can be counted on. You have some work to do to restore your reputation.

What has happened to you? Previously you believed that you wanted to encourage insights. Now you view them with suspicion. It turns out that if you are like most managers, you place a high value on predictability. Your job is much easier if you can accurately predict the work flow, resources, and schedules. Your job is easier if you can accurately gauge progress in moving toward the official project goal. You have fallen into the predictability trap: you are so captured by the lure of predictability that you make it too high a priority.

Insight is the opposite of predictable. Insights are disruptive. They come without warning, take forms that are unexpected, and open up unimagined opportunities. Insights get in the way of progress reviews because they reshape tasks and even revise goals. They carry risks—unseen complications and pitfalls that can get you in trouble. So insights make you work harder. Are you prepared for this additional work? You already have so much to think about and track once your project is under way; you won't have the mental space to anticipate all the implications of the insights that your team members bring you.

We want to boost predictability in order to manage projects and people better. It is a natural pressure for all managers and for all organizations. In 2012, Jennifer Mueller, Shimul Melwani, and Jack Goncalo published a study of why people have an aversion to creativity even though they claim to want creative ideas. The researchers found that if an idea is novel, people automatically assume it isn't practical, reliable, or error free. Novel ideas are associated with failures. Mueller, Melwani, and Goncalo used an Implicit Association Test to pick up unconscious attitudes toward creativity. The participants in the study had to categorize as good or bad a set of objects that were either creative or practical, and the reaction times revealed where they struggled in categorizing creative objects. The study found that creativity was connected in the participants' minds with uncertainty. When people were motivated to reduce uncertainty, they gave lower evaluations to creative ideas. Now we appreciate your skepticism about the novel ideas and insights that your subordinates were bringing you. It stems from your aversion to uncertainty and unpredictability.

If it is any consolation, the Melwani et al. study shows you're not alone. Executives may believe that they want insights and innovations but are most receptive to new ideas that fit with existing practices and maintain predictability. Business organizations treat disruptive insights and innovations with suspicion. Witness the initial hostile reactions to the telephone, to Google's search engine, to VisiCalc, to the Xerox 914 copier, and to Xerox's rejection of its own personal computers. All these innovations became highly successful, but corporations initially were suspicious of these technologies and tried to dismiss them.

To do your job as a manager, you can't depend on insights. You can't schedule them on a timeline. Instead, like all project managers you try to identify all the tasks, schedule start and finish times, and specify the criteria for judging when tasks are completed and when the entire project or program has achieved its goal. Such systematic approaches don't rely on any insights. Your job becomes easier when your subordinates know the tasks, have standards for each task, and can be on the lookout for deviations.

The Perfection Trap

I am defining "perfection" as the absence of errors. Organizations naturally gravitate toward reducing errors. Errors are easy to define, easy to measure, and easy to manage.

The quest for perfection, for mistake-free performance, can be thought of as the War on Error. It is right up there with the quest for predictability. These are both inherent in running an organization that depends on managing people and projects. In well-ordered situations, with clear goals and standards and stable conditions, the pursuit of perfection makes sense. But not when we face complex and chaotic conditions, with standards that keep evolving.

As I revise this chapter in August 2012, I occasionally catch video clips of the London Olympics. Watching Gabby Douglas win the gold medal for all-around gymnastics, I am struck by the scoring. The judges seem to establish a ceiling based on the difficulty of the routine. Then their job is to detect errors—imperfections. Oh, the Russian girl fell off the beam, the commentator says. That will cost her a full point. During the vault Gabby bounces a bit on landing. That's going to cost her. There are very few comments about artistry or exuberance. This scoring method makes the judging easier and more objective. It seems to have changed the sport so that it can be easier to manage. It's another case of defining performance by what can be objectively measured and downplaying the rest. What we come away with is the impression that gold medal performance is all about making the fewest mistakes and perhaps having the biggest smile when landing.

Organizations have lots of reasons to dislike errors: they can pose severe safety risks, they disrupt coordination, they lead to waste, they reduce the chance for project success, they erode the culture, and they can result in lawsuits and bad publicity. In your job as a manager, you find yourself spending most of your time flagging and correcting errors. You are continually checking to see if workers meet their performance standards. If you find deviations, you quickly respond to get everything back on track. It's much easier and less frustrating to manage by reducing errors than to try to boost insights. You know how to spot errors. You don't know how to encourage insights other than hanging inspirational posters on the walls.

Cutting down on errors seems pretty straightforward. You list the steps needed to perform a task, or define the standards for each step, and then check on whether the workers are following the steps and adhering to the standards. You're on the lookout for deviations so that you can bring workers back into compliance. It's like watching a gymnast for wobbles and bobbles.

Perfection can also mean that you carry out the plan exactly as it was drawn up—complete the gymnastic routine just as it was designed. You can buy into that goal. It will get you promoted.

However, insights can take us beyond perfection. They can show us ways to improve on the original plan. Why would we lock ourselves into the initial objective when we can build on discoveries? By this point in your managerial career you know the answer: because there's little payoff for going beyond perfection and lots of downside if you fail to deliver. It's best to settle for perfection. With that, you have fallen into the perfection trap. You are trapped by a desire to reduce errors and achieve the initial objectives. This desire is stronger than wanting to turn out good products and outcomes. You define your job as not making mistakes.

The notion of going beyond perfection strikes some people as odd because they are used to projects with clearly defined goals. When we move into complex settings and work with wicked problems that don't have right answers, we have to discover the goals as we pursue them. The notion of perfection doesn't work in these settings because we don't want to stick with our original vision. We want insights about a better vision than the one we started with. We don't want to be trapped by perfection.

And now we have come full circle to the performance equation that started my exploration of insights (see diagram). The equation shows that to improve performance, we need to do two things. The down arrow is what we want to reduce—errors and uncertainty. My original version referred only to reducing errors, but organizations worry about predictability as well as perfection, so I added uncertainty. The up arrow is what we want to increase—insights. To improve performance, we need to reduce errors and uncertainty and we need to increase insights.

Unfortunately, the two arrows often conflict with each other. The actions we take to reduce errors and uncertainty can get in the way of insights. Therefore, organizations are faced with a balancing act. Too often they become imbalanced and overemphasize the down arrow. They care more about reducing errors and uncertainty than about making discoveries. They fall into the predictability trap and the perfection trap.

THE METHODS

We've seen why organizations would want to act in ways that obstruct insights. Now we'll examine some of the ways that organizations do it. We'll first explore how management controls can interfere with insights. Then we'll see why the very makeup of an organization can block insights.

Enforcing the Down Arrow

The intelligence community provides a good example of an organization that made sweeping changes to try to reduce errors after a fiasco. In 2003, the Central Intelligence Agency endorsed the White House's suspicions that Saddam Hussein was secretly developing nuclear weapons. As a result, the United States led an invasion of Iraq in order to confiscate these weapons, but none were ever found. Consequently, the CIA and other intelligence agencies became even less tolerant of mistakes than before.

Look at the ways that the CIA has tried to cut the chances for errors. It put in place *standards and controls* to reduce the chance of errors. Analysts have to go to greater lengths to carefully *document all their sources, identify their assumptions, and estimate their uncertainty about these assumptions*. Analysts were expected to do these things before, but not to the degree that is now demanded. *Tradecraft*—the best practices of the profession—is defined as taking these kinds of precautions. In some cases, junior analysts are told not to worry about missing anything. As long as they don't violate tradecraft, no one can blame them. So the job is about not making mistakes. One experienced intelligence analyst confided in me that every one of his successes depended on some violation of tradecraft. Now, however, the intelligence community has set its priority on eliminating errors.

The intelligence community has a unit to enforce tradecraft, the Office of Analytical Integrity and Standards. Its job is to reduce mistaken inferences and unsupported conclusions. When I have asked senior officials in the intelligence community if there is a parallel office to encourage insights, all I get are blank stares. Not only is there no such office; the officials have trouble imagining why they would need one.

The British Broadcasting Corporation (BBC) also stumbled over the issue of Iraqi weapons of mass destruction. The BBC was heavily criticized for uncritically broadcasting inaccurate scare stories about Iraq that were spread by a Ministry of Defense employee. Just like the CIA, the BBC tried to protect itself in the future by setting up bureaucratic procedures, including more stringent guidelines and reviews. Managers had to file official forms identifying all possible objections for every program to be broadcast. By 2012, the editorial guidelines ran to 215 pages. The CIA would have been envious.

Then came the consequences of trusting in procedures rather than people. At a crucial moment the experienced BBC staff members and managers who could have exercised judgment were unavailable—either on vacation or laid off in a round of budget cuts. And the procedures weren't enough. First there was an error of omission. The BBC learned that a popular television host who had recently died, Jimmy Savile, had been a child molester. The BBC prepared to report this news, but its layers of procedures got in the way and the report got canceled. Instead, the BBC aired several tributes to Savile's career. Once the child molestation news surfaced, the result was a public uproar directed against the BBC.

Next came an error of commission. The BBC had gotten a juicy tidbit that a member of the Thatcher government had been a pedophile. The BBC ran this story and was again humiliated when it turned out to be false. Apparently no one had checked the credibility of the accuser or given the accused man a chance to respond to the charges. The experienced staff members were unavailable—covering the U.S. elections or pledged to avoid any pedophilia items following the Jimmy Savile debacle—and again the 215 pages of guidelines were inadequate. The people left in the decision making chain were more concerned with following the guidelines than with asking the right questions. They feared that they might be fired if they failed to correctly fill out the forms.

The legal profession is another business that devotes itself to spotting and avoiding pitfalls. A lawyer reviewing a contract is trying to flag potential problems. Lawyers see their job as helping us avoid risks. They get blamed for loopholes in contracts they drew up or reviewed. They don't often get praised if we can use these contracts to accomplish our goals.

Here is a fuller list of common down-arrow methods found in a variety of professions. These are the tools, the weapons, of the down-arrow manager. You can probably add some additional ideas from your own organizations.

Methods for Reducing Errors and Uncertainty
Impose tighter standards.
Increase controls.
Document all sources.
Identify assumptions.
Estimate uncertainty values for these assumptions.
Increase the number of reviews.
Justify conclusions with greater rigor.
Rely on checklists and procedures.
Increase the precision of schedules.

The down-arrow methods are motivated by a desire for predictability and perfection. The methods interfere with insights in several ways.

They are distracting. They chew up people's time on tasks that don't contribute to insights, leaving less time to speculate and imagine. These error-avoidance rituals eat into the time for being thoughtful, the time for spinning stories and following speculations to see where they might go. When analysts have to document and verify and estimate and add caveats to every statement they make, they don't have much attention left over to sort out what they think is really happening.

They make us reluctant to speculate. The pressure to avoid errors can make us distrust any intuitions or insights we cannot validate.

They put insights in a negative light. Anything that smacks of disruption is seen as a threat to schedules. Organizations like to reach closure on decisions and get irritated when asked to revisit them. They close down debate and speculation as soon as possible.

They encourage us to repress anomalies. Anomalies are disruptive, so organizations frequently pretend that they didn't happen. And the more unusual they are, the more we view them with suspicion.

They make us passive. Putting their faith in checklists and procedures, organizations send the message that they just want workers to follow the steps. In such an atmosphere, analysts may shift into a passive stance as they become consumed by the busywork of documenting sources, assigning probability values to uncertainties, and listing assumptions. Analysts may try harder to comply with requirements than to make discoveries. To be fair, most intelligence analysts and other types of knowledge workers still want to make valuable contributions, but their efforts may be more difficult because they're spending so much more time and mental energy on documentation and verification.

A few of these factors popped up in Chapter Ten to explain the contrasting twins. Passivity was there. So was concrete as opposed to playful thinking. Another factor distinguishing the failure from the successful twins was having flawed beliefs. That's the reason for some of the error-reduction strategies—to catch mistaken assumptions and correct them. Good luck. I'm not aware of any evidence that tracking assumptions leads to better performance and fewer mistaken beliefs.

Some of the activities listed in the table have definite benefits: having standards, having management controls, conducting progress reviews, expecting people to justify their conclusions, using checklists, setting up schedules. The problem arises when they are pursued with so much zeal that they interfere with insights.

Take checklists and procedures, for example. Aviation, nuclear power plants, and health care are justifiably intolerant of errors and rely on checklists and procedures to keep error rates low. These checklists are lifesavers that have proven their worth. I wouldn't care to fly with pilots who forget to bring their checklists. I know how valuable the checklists are for protecting pilots from interruptions and memory lapses. I know how valuable checklists are for performing routine functions in nuclear power plants. Similarly in health care, checklists have saved many lives by reducing sources of infection and cutting down on other types of errors.

The downside of checklists is that they intentionally induce *mindlessness*. We just have to follow the steps and not think about them. The important thinking has already been done by the checklist designers. When we follow checklists and procedure manuals, we disengage our active thinking processes, just the reverse of the inquiring mind-set that generates insights.

Checklists and procedures ensure predictability. They work best when all the uncertainties have been resolved and the practitioners converge on a standard way to work safely, efficiently, and effectively. These processes work best in well-ordered situations with clearly understood tasks. Unfortunately, organizations often are tempted to use them in complex situations with ambiguous tasks. A checklist mentality is contrary to a playful, exploratory, curiosity-driven mentality.

I am not implying that we should take errors lightly. After all, the United States did invade Iraq on a futile quest for weapons of mass destruction. However, the job of the intelligence agencies is not only to avoid errors. It is to sound warnings and generate insights. If the CIA becomes too risk averse, then it may not be worth what it costs. Similarly, the primary job of any organization is to produce good products and outcomes, not to avoid errors.

Management controls are necessary for any well-run enterprise to increase predictability and reduce errors. Excessive management controls, however, get in the way of insights because they impose costs—time and effort—for making changes. Performance depends on keeping the arrows in balance.

Organizational Repression

The concept of an "organization" entails a structure of established power relationships through lines of authority and responsibility for assigned duties. Insights are shifts in our understanding, actions, perceptions, goals, and emotions. They are fundamentally *dis-organizing*.

The hierarchical structure of organizations filters insights. Every single level has to sign on if an insight is to make it through to the top decision makers. If junior intelligence analysts detect and announce an anomaly, there's a chance their message will be suppressed somewhere in the chain of higher-ups. This is "organizational repression." I heard of one CIA incident in which a junior analyst concluded that a senior government official in a country was about to lead a coup against its leader. However, the analyst's supervisor refused to recommend this insight for inclusion in the Presidential Daily Brief because it was too unusual. The supervisor didn't want to risk being wrong. The next day, after the White House was caught flat-footed by the coup, the supervisor regretted this decision.

I don't believe this is an isolated incident. Think of FBI Special Agent Kenneth Williams, in Phoenix, Arizona, who spotted suspicious signs two months before the 9/11 attacks. He worried about a coincidence and a contradiction. The coincidence was several Arab men taking flying lessons in the Phoenix area. The contradiction was they didn't want to practice takeoffs and landings, two of the toughest skills to master. Why would someone want to fly yet not care about takeoffs and landings? Williams's Tilt! reflex got triggered. He sent a letter to FBI headquarters on July 10, 2001, the famous "Phoenix memo," warning about a possible terrorist mission. He recommended that the FBI investigate flight schools around the country to see if something similar was happening elsewhere. In addition, he wanted the FBI to sound the alarm with other agencies so that they could coordinate an investigation. He also thought it might be useful to study visa information on foreign applicants to flight schools. However, his warnings were so discrepant, so unusual, that higher-ups in the FBI refused to act on them or pass his warning along.

After 9/11 FBI Director Robert Mueller publicly stated that his organization had no advance knowledge of the attack. His subordinates obviously did an effective job of filtering out the Phoenix memo.

Here's another example, another CIA embarrassment: the agency failed to anticipate that the Berlin Wall was going to come crashing down in 1989 and that East and West Germany would reunify soon after. What went wrong? Jack Davis described the incident in an essay, "Why Do Bad Things Happen to Good Analysts?"

There was an analyst who did notice the shifting conditions. Several months before the Berlin Wall came down, he wrote a draft piece for the CIA explaining that the impediments to German reunification were fading.

Unfortunately, when he coordinated his piece with his peers in the agency, they tweaked and refined it, added caveats, and raised counterarguments. By the end, all the insights had gotten buried. Faint traces were still there for anyone who knew where to look and had seen the original version, but the main insight had gotten stifled by the process of going through a review by experts. As a result, "On November 9, 1989, while CIA experts on Soviet and East German politics were briefing President George H. W. Bush on why the Berlin Wall was not likely to come down any time soon, a National Security Council staff member politely entered the Oval Office and urged the president to turn on his television set—to see both East and West Germans battering away at the Wall."

Chapter Three contrasted Admiral Stark and Admiral Yamamoto, showing how the Japanese could take advantage of an insight, whereas the Americans couldn't. The Americans were overconfident in their power and lacked the organizational willpower to act. The tables were turned a few months later at the Battle of Midway, June 4, 1942. This time it was the Japanese who were overconfident. They had reason for their overconfidence. They had smashed the Americans and British throughout the Pacific—at Pearl Harbor, in the Philippines, and in Southeast Asia. Now they prepared to use an attack on Midway Island to wipe out the remaining few aircraft carriers the Americans had left in the Pacific. The Japanese brought their primary strike force, the force that had struck at Pearl Harbor, with four of their finest aircraft carriers.

The battle didn't go as planned. The Americans had broken enough of the Japanese code to know about the Japanese attack and got their own aircraft carriers into position before the Japanese arrived. The ambush worked. In a five-minute period, the United States sank three of the Japanese aircraft carriers. By the end of the day, it sank the fourth one as well. At the beginning of June 4, the Japanese Navy ruled the Pacific. By that evening, its domination was over, and Japan spent the rest of the war defending against an inevitable defeat.

What interests us here is the war game the Japanese conducted May 1–5 to prepare for Midway. The top naval leaders gathered to play out the plan and see if there were any weaknesses. Yamamoto himself was present. However, the Japanese brushed aside any hint of a weakness. At one point the officer playing the role of the American commander sent his make-believe forces to Midway ahead of the battle to spring an ambush not unlike what actually happened. The admiral refereeing the war game refused to allow it. He argued that it was very unlikely that the Americans would make such an aggressive move. The officer playing the American commander tearfully protested, not only because he wanted to do well in the war game, but also, and more importantly, because he was afraid his colleagues were taking the Americans too lightly. His protests were overruled. With Yamamoto looking on approvingly, the Japanese played the war game to the end and concluded that the plan didn't have any flaws. When we think about how organizational repression filters insights, we should remember the Japanese war game for Midway.

The problem isn't just higher-ups filtering out or rejecting atypical messages; junior analysts often censor themselves. The filtering pervades all levels.

I studied this tendency to suppress anomalies a few years ago in a research project with David Snowden, a leading theoretician in the area of sensemaking—the way people understand sets of messages and data points. Together with Chew Lock Pin and some other colleagues, we studied how individuals and teams detected anomalies. We used a garden path scenario, which is a method for setting people up to have a mistaken view of what is going on and then dribbling in cues to let them discover their mistake. We designed some scenarios that seemed to move logically through a series of events, enticing participants onto a "garden path." Hidden within each scenario, however, was a set of weak signals hinting that something very different was really going on. The weak signals got stronger as the scenario played out. We wanted to see when people would notice the real story and get off the garden path. We studied seven four-man teams, each comprising either military or intelligence officers. The military teams got a scenario about a possible invasion; the intelligence teams got a homeland security scenario. We wanted to find out if the teams would pick up on the actual story early or late.

It wasn't a matter of early or late. None of the teams picked up on the actual story at all. Each team persisted in believing the original story as if nothing had happened.

Then the results took an even more unexpected turn. We had asked each team member to keep a digital diary of his impressions during the scenario. When we subsequently examined these diaries, we learned that in every team at least one person, sometimes two or three team members, noted the weak signals. The individuals on the team had noticed the anomaly, but they rarely brought it to the attention of the leader. In the few cases in which a team member did voice a concern, it got ignored. In other words, these seven teams were less than the sum of their parts. The insights of the individual members never made it to the team level.

When I presented these findings to a U.S. intelligence agency, several junior analysts spoke up to agree with my claims that insights were suppressed. They went further, publicly admitting that they inhibited themselves and that they were reluctant to voice speculations that went against the company line.

Senior analysts in the room were shocked to hear this. They argued that the organization encouraged everyone to give his or her independent views. The junior analysts held firm. What the senior analysts said might be the official position, but it didn't reflect how the organization worked. They didn't want to risk their promotions by making unpopular statements.

As I listened to this dialogue, I was sad to see how blind the senior analysts were to the dilemma that was silencing their subordinates. The senior analysts counted on getting corrective criticisms from the junior analysts, but they were living in a fool's paradise, mistaking silence for assent.

Many organizations claim they are worried about "black swans," the out-of-nowhere crises that can suddenly descend. Black swans obviously threaten predictability. Organizations look for ways to spot these black swans in advance, which misses the point that a black swan is an event that can't be detected in advance. However, in many cases there seems to be a period between the initial manifestation of a black swan and its official recognition and response. The al-Qaeda 9/11 attack is an example; the FBI ignored the early warnings. The financial collapse of 2007–2008 is another example; savvy investors warned of a housing bubble and were ignored.

I suspect that people within many organizations pick up the early signs of a black swan before the organization officially takes notice of it. Rather than trying to do the impossible—spot the black swans in advance—organizations could be better off listening to their alarmists rather than suppressing their warnings. Yet to do so would run counter to organizational DNA, which is programmed to discount anomalies and disruptive insights as long as possible.

There is an irony here. Organizations really are concerned with errors. They don't want to make claims that might be wrong. But in their reluctance to speak out, they miss early warning signals and a chance to head off problems. Their fear of making errors of commission leads to an increase in errors of omission; they block disruptive insights about what is really happening in a situation.

Of course, if every analyst trumpeted each suspicion, that would replace the tyranny of error avoidance with a tyranny of false alarms. Nevertheless, intelligence analysts don't idly proliferate suspicions. The young analysts who worried about the coup and about East-West German unification, the FBI analyst who worried about the bizarre flight lessons, and the Israeli analysts who worried about an Egyptian attack all felt passionately that something was wrong. Their suspicions got squelched *because* they were unusual. That's a sign of risk-averse organizations intent on stifling disruptive insights.

These forces—the need for predictability and the need for perfection (avoidance of errors)—aren't values that organizations *choose*. They seem to be inherent in the very nature of organizations. Even when organizations realize they are moving in the wrong direction, they don't know what to do about it. Some senior officials in the U.S. government intelligence community are worried about all the extra layers of review that have been built in to prevent errors.

One official used the metaphor of a secure password to describe the problem to me. He explained that a four-character password is less secure than one with eight characters, which is less secure than sixteen characters, but is a one-hundred-character password buying much more than one with sixteen characters? At a certain point, the additional workload swamps the small gain in security. Similarly, the additional reviews chew up more energy than they are worth. The additional reviews, the additional effort to reduce errors, creates busywork, consumes time, and crowds out insights.

But even when the officials admit this problem, they don't see easy ways to restore a productive balance between the two arrows. The force of the down arrow is too great.

SCIENTISTS AREN'T MUCH BETTER

If there is any field that should celebrate insights, it is the scientific community of researchers and scholars. Science should exemplify the performance improvement two-arrow equation. Researchers need to reduce sources of error that might contaminate their findings; they also need to increase their insights, the findings from the research they do. Both are essential. The method of science is doubt—to be skeptical and demand credible evidence. The engine of science is curiosity—to speculate and inquire in the service of discovery.

Imagine a continuum of statements, from left to right, with the former absolutely true and the latter absolutely wrong, as in the diagram below.

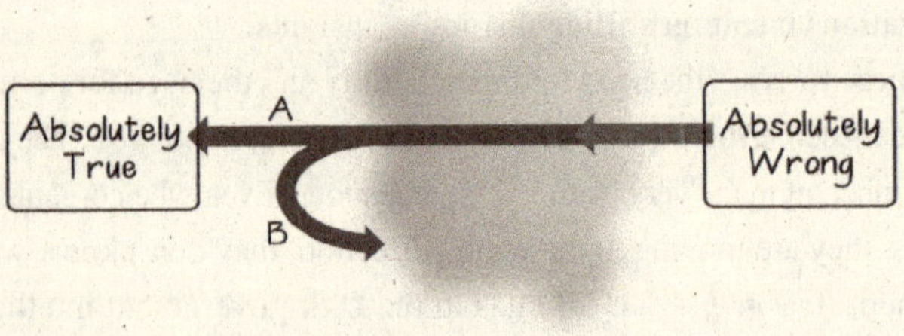

No scientist wants to make statements on the right-hand side. For many, the scientific ideal is to migrate as far away from the right-hand side as possible, all the way to the left. Most researchers want to be known as completely credible. "If she says so, you can take it to the bank." That's high praise in the scientific community.

But what kinds of statements are absolutely true? For the most part, they are vacuous. They carry little information. Scientific progress can slow to a crawl if researchers overemphasize the down arrow, the risk of errors.

Rather than retreating all the way to the left, strategy A in the diagram, scientists, intelligence analysts, and others could use a different strategy: to make the most extreme statements that they can *defend,* strategy B. That means moving away from the safety of the "absolutely true" end point. It means moving toward the fuzzy zone in the middle, getting close to that ambiguous edge. Such a strategy is bound to lead to some errors. If scientists or intelligence analysts make too many errors, that's a problem. But if they never make any errors, they are probably being too cautious.

Organizations like the BBC cannot tolerate errors; their reputation depends on their credibility. Other organizations rightfully fear legal consequences if they make errors. The intelligence community has to worry about its reputation, but it also has to make discoveries and generate insights. Scientists should be even more open to making discoveries than most other specialists, but that isn't always the case.

Thomas Kuhn, the philosopher of science we encountered earlier, noted that scientific researchers tend to filter anomalies just as members of other communities do. He asserted that most scientists spend their careers doing "normal science," basically puzzle solving within the popular research paradigms of the day. As Kuhn put it, "Normal science . . . often suppresses fundamental novelties because they are necessarily subversive of its basic commitments." In the previous chapters we've covered a few cases in which the scientific community tried to suppress ideas: Carlos Finlay's speculations about mosquitoes causing yellow fever, John Snow's suggestion that contaminated water supplies transmitted cholera, Barry Marshall's claim that a bacterium, *H. pylori,* caused ulcers. Scientists aren't any better than the rest of us at resisting the pressure of the down arrow, resisting the lure of predictability and perfection.

Kuhn's description of normal science as "puzzle solving" brings to mind the insight researchers who have spent their careers studying how people solve puzzle problems. As we'll see in the next chapter, they have done their share to shut the gates to insight.

CHAPTER THIRTEEN

How *Not* to Hunt for Insights

LET'S TRY THIS THOUGHT EXPERIMENT. Let's see what we have learned about insights by imagining how we might use the *wrong* tactics to study them. I went hunting for insights in order to assemble my collection of 120 stories. This variety of examples resulted in the Triple Path Model rather than one restricted to a single path.

Now imagine that we are going to repeat this project. What mistakes could we make? What blunders would send us in the wrong direction? What methods would *prevent* us from learning anything valuable? Here are some possibilities:

- Insights pop up unexpectedly, so we'd schedule a specific date, a specific starting and stopping time, for capturing an insight.
- Insights flow from people's own interests, so we would assign an insight task to the subjects we were studying rather than using problems they were already thinking about.
- Insights spring from the themes that matter to us, the themes that keep churning in our minds even when we're not attending to them, so we'd make sure to use a meaningless task that people don't care about.
- Many insights grow over long periods of time, so we'd keep the available time short, no more than an hour.

- Teresa Amabile, a creativity researcher at the Harvard Business School, found that evaluation pressure interferes with insight, so we would make sure people knew we were timing them and appraising their ability.
- A number of investigators, starting with Graham Wallas, have reported that making people verbalize their thought processes can interfere with insight, so we might ask for "think-aloud" reports throughout the session.
- Many insights emerge from connections, coincidences, curiosities, and contradictions, so we would use only tasks that created an impasse and ignore other paths.
- In two-thirds of the cases in my research, people used their experience to make connections and see contradictions, so we'd study the way people handle a task they've never seen before.
- If we wanted to get even more diabolical, we would use methods that make a person's experience actually interfere with insights.

In short, we would design the kinds of experiments currently used in almost all insight experiments. The great majority of studies rely on puzzles presented to college undergraduates under tightly controlled conditions. The researchers may be limiting what they discover about insights because of the methods they use.

Research into insight began in 1917, when Wolfgang Koehler reported how he watched Sultan, a chimpanzee, reel in a banana that was beyond his reach. Sultan tried to use some sticks in his cage, but none were long enough. The chimpanzee sulked for a while, then suddenly fitted two of the sticks together and successfully got his prize. In contrast to the notion of trial-and-error learning, Sultan and some of the other chimpanzees demonstrated sudden insights. Koehler's studies encouraged scientists to study insight using impasse problems that yielded to sudden solutions. College undergraduates replaced chimpanzees because they are more readily available to university researchers and usually require less maintenance.

The nine-dot puzzle is one of the staples of laboratory-based insight research. The subject has to find a way to connect all the dots using no more than four straight lines and never lifting the pencil from paper (see drawing).

The task seems impossible until we realize we can solve it by extending the lines past the border of the dots. The nine-dot problem is typical of insight laboratory tasks. It creates an impasse because we make natural assumptions that turn out to be inappropriate for the task. We assume that we are supposed to stay within the borders. Our kindergarten training to color within the lines kicks in. Two psychology professors, Trina Kershaw, currently at the University of Massachusetts Dartmouth, and Stellan Ohlsson, at the University of Illinois at Chicago, described another assumption that gets in our way: that each turn is supposed to pivot on a dot. Again, our kindergarten games of connecting the dots mislead us. These kinds of assumptions are usually reasonable and helpful in other contexts, just not in this one. We can solve the problem only by discovering the misleading assumptions (see drawing). These tasks produce an "aha" moment of discovery. Up to that moment we have no sense of making progress.

Such tasks fit the creative desperation path from the Triple Path Model of insight. We need a breakthrough, and we consciously examine all our beliefs, assumptions, and anchors to find the one that is getting in the way. Sometimes we get stuck because we've been fixating on traditional ways of using the objects we've been given, and other times the impasse tasks lead us to fixate on inappropriate goals, such as keeping our lines within the "box" of the nine-dot problem.

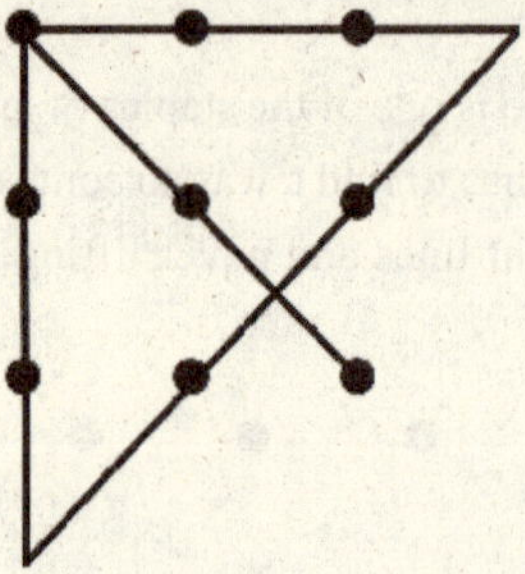

Laboratory experiments on insight travel the creative desperation path. Researchers give the subjects insight problems to solve. The problems lead the subjects to an impasse by setting up an unnecessary assumption. The only way to succeed is to discover this assumption and discard it. Here is another simple example: Two men play five checker games, and each wins an even number of games, with no ties. How is that possible?

Our initial impulse is to examine different combinations of games, but that doesn't get us anywhere. The problem has set up a false assumption: that the men played each other. Nothing in the problem asserts that they were paired, and once we see that we've been trapping ourselves with an unnecessary assumption, the problem is solved. We may groan a little because we feel we've been tricked, and indeed we have. That's what impasse problems do. They trick us into making unnecessary assumptions.

Here are some other common impasse examples. The box-candle puzzle challenges subjects in an experiment to attach three candles to a wall. They are given three small boxes, one containing three candles, the second containing three tacks, and the third containing three matches. Most subjects try to tack the candles to the wall, but this strategy fails miserably. The trick is to tack the boxes to the door and set the candles on them.

The pendulum puzzle asks subjects to discover how to grab hold of two strings that are beyond arm's length. The pendulum problem lends itself to clever manipulations, such as having the researcher "accidentally" brush against one of the strings in order to help the subjects break free of the assumption that the strings need to be grasped while they are at rest. The solution is to discover that a heavy object can be tied to one of the strings and set swinging. A hammer lying on the nearby table can be used, tying it on one of the strings so that it becomes a pendulum bob.

Sometimes real-world situations resemble these tasks. For example, Aron Ralston was stuck, figuratively and literally, until he noticed his arm bend unnaturally when he tried to yank it loose. That hint enabled him to break free of his fixation. He didn't have to cut through the bones of his forearm. He could snap them.

Recently, laboratory impasse tasks have become popular with neuropsychologists who perform brain scans while the subject tries to solve a problem. (A 2010 article by Arne Dietrich and Riam Kanso, researchers at the American University of Beirut, reviews a lot of this research.) The scans show which parts of the brain become activated just before the "aha."

These kinds of puzzles—the nine-dot puzzle, the box-candle puzzle, and the pendulum puzzle—are all *domesticated* insight tasks. The researchers using these tasks aren't doing any hunting or exploring. They are harvesting a predictable crop, one that has emerged from decades of pruning and modifications. Puzzles produce the satisfying "aha" experience we often associate with insights.

I admire the ingenuity of these researchers who delve deeper and deeper into the forces of the impasse and explore different strategies to improve solution rates. Sometimes, however, I have the impression that the strategies for increasing solution rates don't seem to be helpful outside the laboratory. I wish the puzzle researchers would broaden their field of view and look at impasses in natural settings. I wish they would explore the other parts of the Triple Path Model: connection insights and contradiction insights.

I am also disappointed in the way some researchers use the impasse paradigm to disparage the benefits of experience. More than two-thirds of the incidents in my research sample showed people's experience as crucial to insight, but we wouldn't know it from the scientific literature. The impasse paradigm depends on using people's experience against them. The puzzles elicit routine assumptions that turn out to be wrong.

Some researchers deliberately set up conditions that show how experience gets in the way of insight. Consider the water-jar puzzle, a simple arithmetic task for using three water jars of differing capacities to attain the right amount of water in one of the jars (see picture). For example,

Jar A might hold 21 units of water, Jar B might hold 127 units, and Jar C might hold 3 units. Can you measure out 100 units? Yes, by filling up Jar B and then pouring enough out to fill Jar A one time and Jar C two times. The rule here is B minus A minus 2C.

If you present subjects with problems that require them to apply the exact same rule, B minus A minus 2C, you can lull them into a routine. The longer you continue this routine, the more automatic it becomes. At some point you can introduce a problem that has a solution that is more efficient than the routine. For example, you can solve the task of getting 18 units from jars with the capacity of 15, 39, and 3 by using the rule A+C or the original rule (B–A–2C). Control subjects who never got the practice problems are more likely to notice the efficient (A+C) shortcut than "experienced" subjects who have been given several trials that can only be solved using the (B–A–2C) rule. Their so-called experience has blinded them. They have fallen prey to a mental set, called an *Einstellung* effect. The success experiences create an attitude of confidently following a routine, which means that people have fallen into a habit of using an approach that works and fail to scan to see if there might be a better method. These findings are sometimes offered as evidence that the more experience we have, the harder it is to gain insights.

Do we believe that? Do we believe that lulling a person into a mindless routine is the same thing as experience? Do we believe that the more we learn, the more our minds go to sleep? Perhaps that inference makes sense to people who have worked with domesticated insight puzzles for too long. From a hunter's perspective, this critique of experience seems unwarranted. Harry Markopolos used his experience as a fraud investigator to see right through Bernie Madoff. Charles Darwin used his experience on the *Beagle* expedition, his experience observing variations among individuals of the same species, to appreciate the implications of Malthus's ideas. It was the top admirals, Stark and Yamamoto, who grasped the significance of the Battle of Taranto. You can go back through the examples yourself.

I concede that experience can create mindlessness. You have to repeat the same conditions on every trial. Every repetition has to be the same. Researchers call this type of effect "automaticity" because people reach a point where they don't do any thinking anymore. The water-jar study produced an *Einstellung* effect by inducing automaticity. Outside of a laboratory, however, we rarely find work conditions that promote automaticity. When those conditions arise, as on factory floors, the tedious parts of the job are often automated.

The puzzle problems also take advantage of functional fixedness: we have natural ways of using objects, and we don't readily think of unorthodox uses. Boxes contain candles, tacks, and matches, and aren't ordinarily candle stands. Hammers aren't usually pendulum bobs. Yet if we had to approach every situation by thinking of all the possible ways to use each object, our lives might get slowed down a bit. People can be fooled by taking advantage of their usual successful tendencies, but few of the cases in my research sample depended on breaking free of these tendencies. Other than providing grist for the admonition to "think outside the box," these experiments don't seem to have much to teach us.

In my sample, experts did become trapped by their assumptions. Sherman Kent was confident that Khrushchev wouldn't place nuclear missiles in Cuba; Eli Zeira knew that the Egyptians weren't going to launch a surprise attack against Israel. But they weren't trapped because of automaticity or functional fixedness. They weren't trapped by expertise except that their acknowledged expertise gave them too much confidence in their own views. What trapped them was overconfidence in flawed beliefs coupled with reluctance to seriously consider other possibilities. I don't think the water-jar paradigm can shed much light on this kind of trap.

Stellan Ohlsson, one of the leading researchers using the impasse paradigm, has expressed some similar concerns. He has speculated that as far as the impasse paradigm goes, "there is nothing left to explain; the problem of insight can be put aside; no more research is needed on this topic." He has also admitted that the impasse paradigm has failed to consider cases such as Charles Darwin hitting on the idea of natural selection after reading Malthus's ideas on population growth. Ohlsson lamented, "Our current theories are powerless to explain this type of insight."

Research psychologists usually feel uncomfortable with the kind of naturalistic study I did. They prefer to test hypotheses by carefully controlling the conditions, which is why they use the impasse paradigm and perform variations on the same small set of domesticated puzzle problems. They like to define insights in terms of the "aha" experience so that they can have an objective means of distinguishing insight from routine problem solving.

That's where I depart from conventional researchers. I don't believe the purpose of science is to do "good" science. The purpose of science is to learn more about the world, including the world of insights. We don't want to be sloppy about it. We want to use methods that yield results worth taking seriously. We shouldn't, however, become so fixated on the methods that we lose sight of the object of our inquiry. We shouldn't evolve a set of methods that don't fully capture the phenomenon we want to understand.

The insight researchers may have become trapped in these puzzles, trapped by their own assumptions. It may be time to retire the impasse paradigm. We have learned some things from it. But after decades of use it may be played out, like a field that has been farmed for too long. Perhaps it is time to move beyond this paradigm. Perhaps it has become impassé.

PART III

OPENING THE GATES

How Can We Foster Insights?

CHAPTER FOURTEEN

Helping Ourselves

NOW WE ARE READY TO SEARCH FOR WAYS to increase insights. We've examined what happens when people gain insights. We've considered some of the barriers that interfere with insights. Have we learned anything we can use to increase the chances of gaining a useful insight?

In this chapter we'll see if we have learned anything to help ourselves be more insightful. The next chapter will look for ways to help others gain insights. Then we'll tackle the question of what organizations can do to increase insights.

How can we gain more insights? That challenge seems pretty daunting, but we can make it more manageable if we examine each of the insight paths separately. The Triple Path Model of insight suggests the different routes we can take (see diagram).

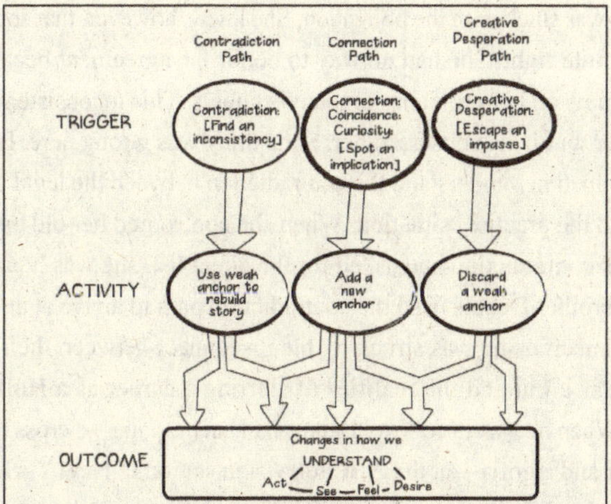

THE TILT! REFLEX

One thing we can do is make better use of the power of contradictions. Think, for example, of detectives trained to notice discrepancies in a person's testimony. On the television series *Columbo,* Peter Falk played a homicide detective with the Los Angeles Police Department. Lieutenant Frank Columbo would seem to be satisfied with the bogus story, which was often told by a sophisticated person who treated the disheveled Columbo as a bumbler who could be easily fooled. Columbo would amble out of the room, pause, turn around, and muse about one thing that was bothering him. That one thing was a contradiction, and it exposed the deception.

We don't need Columbo to spot inconsistencies. We can do it ourselves, using confusions, contradictions, and conflicts as springboards to insights. Usually we feel frustrated when we run into these kinds of disruptions, but they represent openings for making discoveries. We just have to replace our feelings of consternation with curiosity. We can take advantage of our Tilt! reflex.

Think back to the example of Ginger, the young woman who was trapped by a noncompete agreement once she changed jobs. On the surface she was stuck with the obligation. She knew, however, that something wasn't quite right. She had no way to honor the agreement because she didn't know all her previous company's clients. This inconsistency frustrated her, but it also obsessed her. Something was wrong here. Her Tilt! reflex helped her understand the contradiction between the legal requirement and the practical situation. When she confronted her old employer with this contradiction and asked for the client list, she was home free.

My brother Dennis used the contradiction path to arrive at an insight about himself as he was struck by the disconnect between the routines he used as a kid and the realities of starting a career as a Hollywood writer. When he was in college, Dennis had been a page—a cross between an usher and a gofer—at the ABC television station in New York. When he moved to Los Angeles, he applied for a similar page job at an ABC station, and ABC said it intended to hire him. Dennis was excited to hear that, because a network page job can be a stepping-stone to a show business career. It can be a way to meet high-ranking entertainment industry contacts.

Weeks went by and ABC never called him, so Dennis sulked and crossed the network off his list of possible jobs. Then Dennis got a call from the person who would be his manager at ABC offering him the page position. Dennis went into a litany of complaints: "Well, I waited; you never called; how was I supposed to know?" etc., etc. Dennis went into the same routine he had used growing up in the Bronx to assign blame.

The manager cut him off. "Look, do you want the job or not?"

Dennis paused for an instant and then replied that he did want the job. Later, looking back, he realized that that was the instant—Tilt!—when he began to grow up. He suddenly appreciated that no one outside his immediate family cared who was really to blame. People wanted to get work done. What counted was whether he could deliver and be trusted, not whether he could avoid responsibility or justify accusations. The games that Dennis had mastered within our family weren't being played outside. It was time to leave them behind. Although not completely behind, because Dennis did go on to create HBO's *The Larry Sanders Show* with Garry Shandling, which was a comical version of the blame syndrome he had lived through as a child.

Dennis's epiphany helped him make the transition from a displaced postadolescent to a successful adult. His insight concerned the games he needed to abandon to build a successful career. The insight also was about what matters to effective managers. Dennis would have to balance his needs for self-worth and entitlement against the pressure of working collaboratively in a large studio.

Ginger and Dennis illustrate how people use contradictions rather than deflecting them. The next example describes how a fire department captain deliberately opened himself up to a contradiction and learned more than he expected. His experience shows how contradictions can inform and transform us.

In 1996, my colleague Caroline Zsambok and I ran a three-part workshop on On-the-Job Training (OJT) for the Los Angeles County Fire Department. Many people believe they learn their important work skills while doing their jobs and not in formal training programs or classrooms. Our workshop taught the training officers how to use OJT to speed up what firefighters learned from their experiences.

In the first session, one of the training officers, a fire captain, asked if our OJT approach applied only to skills or if he could also use it with firefighters who had attitude problems. Caroline and I told him that the workshop wasn't designed to address such issues. We were not psychotherapists. (Perhaps because we were introduced as psychologists, he had jumped to that conclusion.) We couldn't address personality or attitude problems.

When we came back a month later to put on part two of the workshop, the same fire captain raised his hand and told us we were wrong. After the first session, he had gone back to his unit and to business-as-usual. When his unit went out on a medium-sized fire, he gave an assignment to one of the young firefighters—the one he had in mind when he had asked about attitude problems. A month earlier he had firmly believed that this firefighter had an attitude problem.

On this occasion, as usual, the firefighter didn't do what the captain expected. After the crew returned to the station, the captain called the firefighter into his office, ready to chew him out once again, tell him that he was incompetent, and announce that this incident would be written up and added to his already thick file, documenting why he should be drummed out of the service. But just as the captain was getting ready to begin his tirade, he glanced over to the OJT poster we had provided and decided to start the discussion in a new way. The poster listed some of our OJT tenets. One of the tenets stated there were other ways to train people besides telling them what to do, such as starting off by asking trainees to describe what they were trying to do. The captain figured he would give this approach a shot and then ream the young firefighter out. So he began in the way we recommended, by asking the firefighter what he had in mind in the way he carried out the assignment.

To the captain's surprise, the young firefighter actually had a very good rationale for his actions. He was not being deliberately provocative or insubordinate. He was just doing the best he could. "And then I realized," the captain told our class, "that *he* wasn't the attitude problem. *I was the attitude problem*." Tilt! The young firefighter had his own valid plan, and without questioning him, the captain would never have learned it. The captain gained an insight into his young firefighter and a self-insight that he was creating unnecessary conflict by getting angry at every departure from the rigid procedures he had demanded. The captain gained his insight because he sought out someone else's perspective and opened himself up to contradictory views.

SWIRL

The connection path thrives on having lots of ideas swirling around and on making accidental linkages. The more swirl and turbulence, the greater the chance for a discovery. Martin Chalfie walks into a lunchtime seminar on jellyfish. Michael Gottlieb gets a referral with bizarre symptoms. Jocelyn Bell Burnell spots some unusual squiggles in her radio telescope recording. Barry Marshall decides to satisfy a research requirement by studying a corkscrew-shaped bacterium. A firefighter sits in an OJT seminar and learns a method that helps him resolve an attitude problem.

If we increase these accidental linkages, we might be able to gain more insights by expanding our exposure to unfamiliar activities. We can try to have more varied experiences ourselves, or we can increase our encounters with different kinds of people, working in a variety of areas, peppering us with new ideas. Each of these ideas could become a new anchor for us, seeding a new combination of concepts.

In his 2010 book, *Where Good Ideas Come From,* Steven Johnson recommends ways to increase creative turbulence. According to Johnson, we should find ways to increase the density of ideas to which we are exposed and to increase our contact with creative people. We should foster serendipity—the random collision of ideas. We should increase our intersections with different communities, using real and virtual gathering places such as coffeehouses and social networks. We should encourage group and network collaboration as opposed to individual efforts. We should take on multiple hobbies. Each of these recommendations would strengthen our chances of making unexpected connections.

Some organizations try to increase internal swirl by devising schemes to force their knowledge workers to encounter colleagues from different specialty areas. One way is to randomly assign offices instead of letting each specialty clump together in its own ghetto. When he designed the new Pixar building, Steve Jobs tried to increase swirl by promoting face-to-face encounters in a central atrium. He even went as far as planning only two bathrooms (gender segregated) in the entire building, both connected to the atrium. A revolt ensued, initiated by a pregnant woman who objected to a ten- to fifteen-minute walk to the bathroom, and Jobs reluctantly sprinkled in some additional bathrooms.

The connection path is where the advice to increase swirl and turbulence makes the most sense as a way to stumble on a worthwhile combination of ideas. Here is where the creativity methods for promoting random combinations come in. They could help us find new blends we otherwise would never have considered.

Despite the popularity of this strategy, I am not very enthusiastic about methods to increase swirl and turbulence to create more chances for accidental discoveries. Stories about insights that emerged accidentally illustrate the value of swirl, but we can't just reverse the process or we run into the fallacy of backward thinking: if we identify some conditions that accompanied good ideas in the past, then in the future we just have to create those conditions and we'll get more good ideas. Perhaps. But we'll get lots and lots of bad ideas as well. Recall Martin Chalfie's lunchtime revelation when he heard about the green fluorescent protein and Alison Gopnik's discovery when she listened to her son commenting on the dessert she was making. We can't create more Chalfie moments by tripling the lunchtime seminars. We can't create more Gopnik discoveries by preparing more exotic desserts or transcribing all the comments of two-year-olds in accordance with the adage "out of the mouths of babes."

Many people spend large chunks of time in coffeehouses and never get more than a caffeine buzz.

The more random combinations we produce, the greater the burden to screen out the useless ones. Insight involves making a new discovery without having to consider bad ideas. As the mathematician Henri Poincaré put it, "Creation . . . does not consist in making new combinations. . . . The combinations so made would be infinite in number and most of them absolutely without interest. To create consists precisely in not making useless combinations."

CRITICAL THINKING

The path of creative desperation calls for a different stance. We're cornered and must find a way to escape a trap set in our own minds. We're overlooking something. We're making an unwarranted assumption.

Some error-phobic organizations encourage their knowledge workers to list all their assumptions, the better to ferret out the bad ones. I have never seen any benefit from such an exercise, and I'm unaware of evidence that assumption-listing has any value.

I suspect the strategy of listing all assumptions stems from the fallacy of backward thinking we just discussed. The rationale is that if you trace insights backward to their origins and find that many depend on rooting out questionable assumptions, then you can improve performance by having people keep track of all their assumptions so that they can identify the weakest ones.

The problem, though, is that people make lots of assumptions when confronted with challenging situations, so the task of listing all of them could be very distracting and tiresome. Further, we need expertise to judge which assumptions are weak. The battle of Toulon illustrates the kind of background—the generally prepared mind—that Napoleon needed to reexamine the tactical situation. If we ran the critical thinking exercise with other French army officers, asking them to list assumptions, it is unlikely that they would produce the insight about shifting attention away from Toulon and toward the forts guarding the harbor.

There is no evidence that Napoleon or any of the decision makers in this chapter went through a process of systematically listing and evaluating all their assumptions. I'm not aware of any evidence that such an exercise increases insights in natural settings, even if people have the time to carry it out.

What we need is a type of divining rod that directs us to the shaky assumptions we've been making. For example, if we are part of a team doing a project, perhaps we can recall a conflict among team members about what was really happening, or a time when someone who was usually competent became confused about something. Perhaps we can think of a surprise, something we expected that didn't happen. Or a contradiction. Any of these difficulties might offer clues about our flawed assumptions.

Probably the most effective strategy is to engage in something called "critical thinking," a systematic analysis of evidence, a process of thinking that questions assumptions, not by asking people to exhaustively list all their assumptions, but by logically reviewing the evidence in order to decide whether or not to believe specific claims.

As long as we keep critical thinking in perspective, it is clearly an essential skill if we want to make good judgments and decisions. However, some critical thinking advocates seem to go overboard with it. Like any skill, it has boundary conditions; there are times to ease back on it. Too many risk-averse organizations have embraced critical thinking programs and methods as a way to reduce decision errors without thinking about how critical thinking can dampen the up arrow and interfere with the noncritical, playful attitudes we need to explore new ideas. In many ways, thinking critically runs counter to forming insights. Still, if there is any place for critical thinking in fostering insights, it would be in times of desperation when we need to undo the flawed assumption that is getting in our way.

THE PAUSE THAT REFRESHES

What about the value of incubation for fostering the flash of illumination? Wallas enshrined incubation in his four-stage model, and many other scientists have endorsed the importance of incubation in their work.

Only 5 of the cases in my research sample exhibited an incubation process versus 47 incidents that happened too quickly for incubation—think of Wagner Dodge running for his life, Chalfie in the seminar, Markopolos seeing Madoff's investment materials, Gopnik listening to her son's complaint about the dessert she was making, and Aron Ralston noticing he could snap the bones in his arm. So incubation isn't necessary for insights to occur. The remaining 68 cases were too difficult to code one way or the other.

David Perkins, in his book *Archimedes' Bathtub* (referring to the story about the famous Greek philosopher who, while taking a bath, had an insight about measuring the volume of gold in the king's crown and ran naked down the street shouting, "Eureka!"), asserts that scientific research has failed to demonstrate any incubation effect. In experiments in which subjects are made to wrestle with impasse problems, some get a bit of time off, whereas others don't. It doesn't make any difference. The time off doesn't help the subjects solve problems.

Perkins also admits that the controlled conditions may have prevented incubation from having its full effect. And now I will make my own admission. While my critical thinking makes me skeptical of incubation, emotionally, perhaps superstitiously, I am drawn to it. I try to review my notes on a project before taking off on a long bike ride. During a lengthy writing project I will look over my outline for the next chapter before going to sleep in order to give my unconscious mind a chance to get to work.

I was therefore encouraged to find a paper by Rebecca Dodds and her colleagues in the Psychology Department at Texas A&M reviewing experiments that show incubation increasing our chances of having an insight. Incubation has a stronger effect when researchers increase the incubation period, and also when the person prepares more thoroughly prior to incubation—just as Wallas claimed. This review identified 39 experiments on incubation, dating from 1938. Roughly two-thirds of the experiments, 26 out of 39, found an incubation effect. These experiments used controlled tasks such as impasse problems, so it isn't clear if the findings will generalize to everyday settings. Still, I was heartened by their findings.

How could incubation promote insights? In a 2010 review article about incubation, insight, and creative problem solving, two psychology researchers, Sébastien Hélie, at the University of California, Santa Barbara, and Ron Sun, at Rensselaer Polytechnic Institute, listed some possibilities. One idea, dating all the way back to Wallas, is that during incubation our unconscious mind continues to chew on the problem. A second idea is that incubation may work by letting us recover from mental fatigue. Prolonged concentration may drain our mental energy, and so a period of relaxation may be needed to get us going again. A third explanation is that in a relaxed state we can summon forth remote associations that might otherwise be blocked by our critical analyses. We can playfully daydream about the problem, exploring uncommon connections. A fourth view is that incubation may let us build on chance events. After working hard on a problem, we are primed to notice implications in all kinds of places. Even though we aren't deliberately working on the problem, we are in a state of heightened sensitivity about it. The Alison Gopnik case fits here. There she is, making the dessert, keeping an eye on her son, and whammo, she sees how she can study the minds of infants.

Incubation should be most useful when we're on the creative desperation path. I'm tempted to say that incubation is least relevant to the contradiction path, but I can remember times that an inconsistency popped into my head hours or even days after I got exposed to the inconsistent data points, so scratch that suggestion. I also wonder if the different mechanisms listed by Hélie and Sun might each apply to different insight paths. For instance, we need to recover from mental fatigue only if we've been wrestling with an impasse, the desperation path. We need to summon remote associations only on the connection path.

When it comes to ways for gaining insights, the Triple Path Model describes different routes we might take. We use the contradiction path to notice, search for, and apply inconsistencies and anomalies. We use the connection path to increase our exposure to novel ideas. When we get stuck, we use critical thinking methods to locate and correct flawed assumptions and beliefs.

Now we can cut through some confusion about ways to achieve insights. Is it important to be exposed to lots of different ideas? Perhaps for those using the connection path, but not for the others. Is the insight process a matter of breaking through fixation and impasses? For situations involving the desperation path, but not for the others.

CHAPTER FIFTEEN

Helping Others

OUR CHALLENGE IS VERY DIFFERENT when we shift from helping ourselves to helping others. It often means trying to correct their flawed beliefs, which in turn means we have to understand what these flawed beliefs are. The fundamental part, often the toughest part of helping others, is diagnosing the confusion—determining what is wrong with their thinking.

DIAGNOSIS

My daughter Devorah, when she isn't playing fantasy baseball (she's been the commissioner of our league for the past decade), is a cognitive psychologist. Devorah works on designing new products, trying to understand how the people who would use them might fit them into their lives. One of her projects involved ways for people with impaired vision to use an adapted Kindle-type reader. A couple of years ago she was working with a seventy-nine-year-old man, a retired English professor, who was frustrated because his poor vision was preventing him from reading books. Devorah showed him how the adapted Kindle type of device could let him continue to read. Though completely computer illiterate, he followed her demonstration, but he was still very dubious about the device. Something was obviously bothering him, but Devorah didn't know what it was. She was patient but persistent and got him to the point of being intrigued about the possibilities the device offered for enabling him to once again experience the pleasure of reading. Not enthusiastic, just intrigued. Then he became strangely quiet.

Instead of moving on to the next phase of the interview, Devorah waited for him to complete his thoughts. After a short while, he said that he could see how this technology could be useful to him. But he would need to clear off a shelf in his apartment to make room for the devices.

His comments were a revelation to Devorah. She suddenly appreciated why he was so nervous about the technology. She diagnosed his flawed belief: He thought the electronic readers were like Books on Tape, an electronic version of the audiocassettes or CDs he must have been listening to, each holding a single book. Devorah felt an insight rush. She also felt a delicious surge of power because she was about to tell him something that would jolt his thinking. She savored those feelings for a few seconds and then said to him, "You know, you can store more than one book on each device." "How many?" he asked. "Dozens, perhaps hundreds." "Whoa!" he exclaimed, throwing out his arms, "That's amazing." Devorah later told me that she could feel his brain rewiring at that instant.

Notice that Devorah waited, even though the man wasn't saying anything, because she could see he was thinking. She didn't know what he was mulling over. She just sensed that he needed some time to sort out his reactions. And instead of blurting out the capacity of the device, she provoked his curiosity so that he posed the question to her. She didn't rush to tell him how many books he could store on each reader.

Helping people correct their flawed beliefs doesn't mean offering unsolicited advice. We don't always have to tell other people what we think they need to know, which usually entails trying to convince them that they need to know it. We know how tedious and irritating that can be. It's much better to be patient and plug in to the person's desire for insights rather than trying to push information.

Psychotherapists have the advantage that clients come to them hoping for insights. They specialize in helping their clients achieve insights about why they make poor decisions and why they repeat the same destructive patterns. Much of the business of psychotherapy revolves around the insights the therapist gains about the clients and the insights the clients attain about themselves.

My brother Mitchell, a psychotherapist with over thirty years of experience, told me about one of his cases, a woman we'll call Barbara who was running a business with her cousin. Barbara came to Mitchell because she was anxious and depressed, largely due to frustration with her cousin's irresponsible behavior. The cousin refused to hold up her end of the partnership. Barbara didn't understand her cousin's actions and why she was being so cruel. She couldn't think of anything she'd done to provoke her cousin's anger, but she must be missing something. Barbara wondered what was wrong with her. Her cousin was always so much fun to be around; that's why she had jumped at the chance to go into business with her. And now it had turned into a nightmare.

Mitchell listened to the complaints and even met with the cousin, and he concluded that the problem wasn't with the cousin's actions or anger but rather with her extreme narcissistic personality. Barbara's cousin treated social interactions as a one-way street and demanded all the attention, support, and recognition. Her excessive self-absorption took center stage over harmonious relationships and even over the needs of the new business.

Mitchell then had to help Barbara understand her cousin's narcissism and correct her belief that her cousin was deliberately trying to hurt her. He described the patterns he saw and discussed the nature of narcissistic personality disorder. He even provided Barbara with some material to read.

Barbara came back the next week amazed at what she'd learned and how it lined up with her cousin's actions. "That's it," she said. She felt freed from the quagmire of her relationship and freed from her anger at her cousin. She no longer thought her cousin was betraying her or was trying to deliberately hurt her. She even had some sympathy for her cousin's condition. Nevertheless, she had to extricate herself from the partnership. It was never going to work. Once Barbara took over the company and forced her cousin out, the business started going gangbusters. She could run it as a business, not as a vehicle for her cousin's insatiable needs. In their last session, Barbara told Mitchell that finally she was very much at peace.

Both of these stories illustrate the important first step of diagnosis for helping others achieve insight. Once Devorah and Mitchell discovered the flaw in the person's thinking, it was easy to provide the information that would set things straight.

There are times when it isn't so easy. Times when insight alone isn't enough. Times when we have to facilitate insight and then translate it into action.

DIAGNOSIS PLUS ACTION

Just because we have a good insight doesn't mean we'll behave with more maturity or make wiser choices. That's why psychotherapists don't get too excited when their clients achieve self-insights. The therapists want to see these insights translated into action, the way Barbara took over the business she'd been running with her cousin. Sometimes the clients never change their behavior. Other times, the insight is the starting point for clients to better manage their emotions and their behaviors.

My friend Bob Barcus, a highly experienced psychotherapist, told me about a case in which a mother and her depressed teenage daughter came to see him about their strained relationship. The mother, who seemed a pleasant and competent woman, had nothing good to say about her daughter. Bob's impression of the daughter was that she seemed to be a good-natured, sweet kid who wanted to regain her mother's love and approval. Bob decided to work on the relationship rather than on the girl's depression. And instead of trying to ferret out what had gotten their relationship so tangled up, Bob chose to take steps to untangle it.

Bob usually doesn't like to make any interpretations or suggest any homework during the initial session, but after watching and listening to them, he asked the mother to tally the number of positive and negative comments she made to her daughter during the next week. He didn't ask her to change anything. Just to keep score.

The following week Bob met with the mother and daughter separately. When he asked the mother about the tally, she burst into tears. "I was all negative." She was crushed by how consistently critical her comments had been. So Bob's suggestion worked. His client gained an important insight about the way she was treating her daughter. This insight alone, however, was unlikely to accomplish anything except cause the woman pain. The mother had nowhere to go with it. Yet the insight, and her emotional discomfort, did provide a powerful motivation to change.

Bob made another suggestion: He asked the woman to reverse the negativity by noticing and praising her daughter's positive behaviors. The mother resisted; she feared that her daughter's sullen anger would prevent any behaviors to praise, so Bob advised his client to look for the absence of negative behaviors in her daughter and praise these. For example, if the girl went for even five minutes without disrupting the household harmony, the mother could express appreciation for the peaceful time.

At the third session, the mother was beaming and so was the girl. They were laughing and joking again. Their home had stopped being a battleground, and the daughter no longer seemed depressed. For the first time the mother felt she could reclaim her child. A follow-up a month later showed that they were still on course.

This story shows that helping others can be more complicated than correcting their beliefs. To be effective, we may have to guide the person into new ways of behaving. The next incident shows one of my rare successes.

Several years ago, my friend Jay Rothman invited me to sit in on a workshop he was running so that I could observe methods he was using to help people set goals. One of his exercises was to have participants think of a word that described something they cared deeply about, along with a story to illustrate that word. Even though I was just an observer, I decided to join the exercise. But that meant coming up with a word and a story.

I couldn't think of any words, but there was one story that took center stage in my mind. It was the time my friend "Jimmy" visited me and made an unusual request. He asked if I would take him to play racquetball. He knew that I played a lot, but he had never before shown an interest in the sport. Seeing that his request surprised me, Jimmy explained that his new girlfriend was an avid player. When they played together, she wiped the floor with him. So he just wanted to give her a good match. He didn't need to win. He just needed to stop humiliating himself.

Jimmy and I got on the court and warmed up for a bit, just hitting the ball around softly. Then I said we should play a game. I let Jimmy serve first. He lost the point, and it was my turn to serve. I served a moderately paced ball to his backhand. He made an awkward lunge as it came near and he missed. I tried another serve to his backhand. Same result.

At that point, I said the game was over. Instead, I was going to work on his backhand return of serve. I was going to serve a series of balls to his backhand—but I didn't want him to swing at any of them! I just wanted him to watch how the ball bounced off the back and side walls.

My reasoning was that Jimmy seemed so anxious to lunge at the ball that he probably had never gained a sense of the bounce trajectories—how a ball comes off the back wall, what happens if it kisses the side wall first, and so forth. He wasn't going to learn through experience because he wasn't giving himself a chance to have useful experiences. That was my diagnosis. It explained Jimmy's problems, and it also suggested that continued practice wasn't going to get us anywhere.

We tried this new drill for a while, me serving and Jimmy watching how the ball came off the side and back walls. Then I added a wrinkle. I continued to serve to his backhand. He continued not to swing. But he would say "there" and point to the ball each time it caromed to a place where he would have a good swing at it. I wanted Jimmy to watch the ball more actively and learn all the different kinds of shots he could make.

After we did that for a bit, Jimmy asked if he could start swinging at the ball. Previously, he had been nervous at having to hit a backhand, but now he was eager. I told him that was okay but that he shouldn't feel he needed to. He started hitting good backhands, and we finished the session with a lot of enthusiasm.

Jimmy went on to play well against his girlfriend. Soon he was winning most of the games, at which point she lost interest in playing racquetball with him. They eventually broke up, but Jimmy has assured me it wasn't because of his new racquetball prowess.

I like this story because I didn't tell Jimmy anything. All I did was arrange for him to make his own discoveries. When I told the story to Jay's workshop, the word that accompanied it was "discovery." The exercise helped me realize how much I enjoy helping others make their own discoveries.

None of the examples we've covered thus far require anyone to abandon closely held beliefs. That rejiggering of beliefs takes more extreme effort because the people we want to help will resist giving up the beliefs anchoring their understandings. Getting them to change takes a skillful use of contradictions.

USING CONTRADICTIONS TO REPAIR FLAWED BELIEFS

The cases that impressed me the most were ones in which people wielded contradictions like magic wands, creating constructive dilemmas. One of my favorite examples is the time Doug Harrington went from being a master pilot to an incompetent pilot and back again in the space of a single day. Harrington was a skilled navy pilot and had been flying for about twelve years, most recently in F-4s. He was an instructor pilot, teaching younger pilots how to fly and especially how to land on aircraft carriers, a very demanding skill.

Harrington had just been chosen to transition from flying F-4s to flying a different airplane, an A-6, with more sophisticated weapons systems. He quickly mastered the A-6. All he had left to qualify as an A-6 pilot was to do his carrier landings. He was scheduled to do six daytime landings and then four more that night. But it didn't work out.

Harrington was coming in for his first daytime landing. He thought he was perfectly lined up, but the landing signal officer (LSO), who was responsible for guiding pilots to safe landings, radioed him, "Come right, come right." This message made no sense. Harrington at first ignored this directive, but the LSO was insistent, so Harrington tried to comply but did so too little and too late to satisfy the LSO, who waved him off. The LSO's behavior was bizarre. Harrington flew around for another try, lined up perfectly, and was again told, "Come right, come right." He did better this time and landed, but it was a ragged landing. Harrington butchered all his landings that day. He was told he wouldn't be allowed to try any nighttime landings. He would have to repeat the daytime landings the next day. If he didn't do better, his flying career in the navy would be over.

He spent the rest of that evening brooding because he had no idea what had gone wrong. Each landing had been correctly lined up, and each time the LSO had insisted he come right, and his attempts to comply had been clumsy. His landings were ugly, which was bad enough, but he didn't know why and he had no reason to expect things to get better.

That night, the senior LSO visited him and asked what his strategy was for landing the airplane. Harrington explained that he tried to put the nose of the plane on the centerline of the runway, the strategy he always used.

The senior LSO questioned him further and learned that Harrington had been flying F-4s, an airplane that placed the pilot in front of the navigator. When Harrington flew the training version, the student pilot sat in front of him so that both of them were aligned in a straight line from the nose of the plane to the tail. Now in an A-6, Harrington was sitting next to an instructor, slightly to the left of a line from nose to tail.

The LSO thought he saw the problem: Harrington's strategy of putting the nose of the plane on the centerline of the runway wouldn't work if he were offset from the airplane's centerline. The LSO explained his diagnosis to Harrington, who rejected it. Harrington argued that he was only about a foot and a half off the centerline, and therefore it shouldn't make any difference.

So the senior LSO devised an exercise on the spot. You should do this exercise as you read along. The LSO asked Harrington to hold his arm straight out with his thumb up. The thumb represented the nose of the airplane. Harrington next closed one eye and aligned his thumb against a vertical line someplace in the room, such as the door frame. Go ahead and do that. The vertical line represents the runway on the aircraft carrier. Then Harrington was told to move his head one-and-a-half feet to the left, to reflect the fact that he was not sitting on the centerline, and then realign his thumb against that same vertical line. When you realign your own thumb, you will need to pull it sharply over to the left.

In doing this exercise, Harrington discovered the flaw in his thinking. He had just pulled his thumb, the nose of his make-believe airplane, sharply over to the left of the runway. He immediately knew why the LSO had been telling him to "come right, come right." The next day he nailed all his landings.

Look at how the LSO helped Harrington. The LSO knew something was wrong, and even though he had no responsibility for the pilots' performances, he visited Harrington and initiated the discussion. The LSO didn't have anything to tell Harrington. He didn't give him any lectures. He was there to diagnose the problem. Once he spotted the flaw in Harrington's thinking, the LSO quickly devised an exercise enabling Harrington to discover the flaw for himself. The exercise created an experience that directly contradicted Harrington's belief that he merely needed to put the nose of the plane on the centerline. The contradiction forced Harrington to abandon his core belief about how to do landings. The insight was both conceptual and emotional. As Harrington told me the story, his face re-created that instant when he pulled his thumb to the left and then winced as he realized his stupidity. The LSO in this story exemplifies the ideal teacher/coach/advocate.

I thought of Harrington when I read an article about a master teacher named Deborah Ball who was teaching arithmetic to her third-grade class. She was teaching the students to distinguish odd and even numbers. A student named Sean volunteered his discovery that the number "6" could be odd as well as even. He proceeded to give his reason: three 2s make 6. So 6 is composed of an odd number (3) and an even number (2) and therefore is both odd and even.

What happened next is what I find so striking. Deborah Ball didn't try to correct him, nor did she ignore him and move on with the lesson. Instead, she decided to let Sean and the rest of the class figure it out. She let Sean explain his confused reasoning. Then she asked the class to comment. The class joined in a discussion of the attributes of odd versus even numbers. Another student came up with a definition that was easy to remember and, more importantly, easy to apply. An odd number has one left over if it is grouped by twos. It was a rule that contradicted Sean's claim that "6" could be both odd and even. Even Sean accepted that rule.

The class also identified a category that blended odd and even numbers. A student, Mei, said the number "10" had the same properties that Sean's number 6 did: 2 times 5. This new category consisted of even numbers that are made up of an odd number of groups of two. The class called these "Sean numbers."

The point here is that the teacher didn't correct Sean. Instead, she, the class, and Sean himself examined the flaw in his reasoning, and by the end of the class, Sean self-corrected his thinking just as Harrington had. Sean had an insight into odd and even numbers because he corrected his own flawed belief.

Ball saw the flaw in Sean's thinking, but she didn't rush to help him. She instead trusted her class, and, with her guidance and encouragement, the students not only articulated a clear rule for identifying odd and even numbers, but also invented the new category of Sean numbers. The rule for differentiating odd and even numbers contradicted Sean's claim, just as the LSO's thumbs-up demonstration contradicted Harrington's landing strategy. Both cases illustrate the power of contradiction to repair flawed beliefs.

Deborah Ball was not obsessed by the right answer, nor was she troubled by the error Sean introduced into the discussion. She wanted to get inside Sean's head. As Sean explained his reasoning, Deborah Ball could see where he was going wrong. By appreciating *why* a student like Sean might get a wrong answer—by unpacking the erroneous beliefs that were misleading Sean—she could formulate a strategy to help him and others in the class like him correct their beliefs. As she explained later, "Teaching depends on what other people think, not what you think."

This kind of teaching is very hard. It takes more than knowing the right answers. It takes curiosity and compassion and the ability to decenter—to take someone else's perspective. It depends on skill at using contradictions to help people make discoveries. It is an ability that I admire and envy.

I still pursue my goal of trying to help others gain insights, the goal I described in Jay Rothman's workshop. I have a long way to go. In the next example, I found a way I could have helped a marine gain an insight about his mental model of ambushes. It took me only fifteen years to figure this out. I would like to shorten that reaction time when I encounter similar problems in the future.

FUMBLING

I was working on a project to train decision making skills in young marines, sergeants and lance corporals, as they prepared for an exercise in 1997. During the first training session, I presented a simple squad-level paper-and-pencil decision game. The platoon had three fire teams and was heading left to right on a path; there were two hills just north of the path, one on the left and a slightly larger one on the right. It was late in the day, dusk.

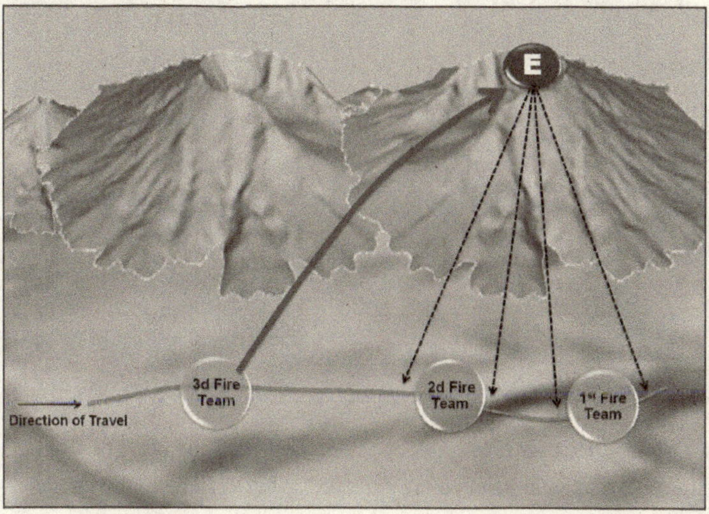

The enemy was located at the top of the hill on the right, the position marked "E" (see drawing), and had just conducted an ambush that pinned down the first two fire teams. These marines had taken cover as best they could, but they were pretty helpless. The decision maker in the exercise played the role of the leader of the 3rd fire team, which had not yet entered the ambush zone. "Quick," I said to the muscular sergeant who was on the hot seat. "What will you do? You're the leader of the 3rd fire team. Issue the order to your team."

Without any hesitation he gave his order: "Follow me, men," as he showed how he would lead his fire team up the hill on the right, charging directly at the enemy position. This response struck me as a really dumb idea, but I didn't know how to say that, so I asked other noncommissioned officers (NCOs) in the room what they would do. Their reactions were more in keeping with my thoughts—move up the hill on the left and use that position to drive the enemy off and rescue the marines who were pinned down (see second drawing).

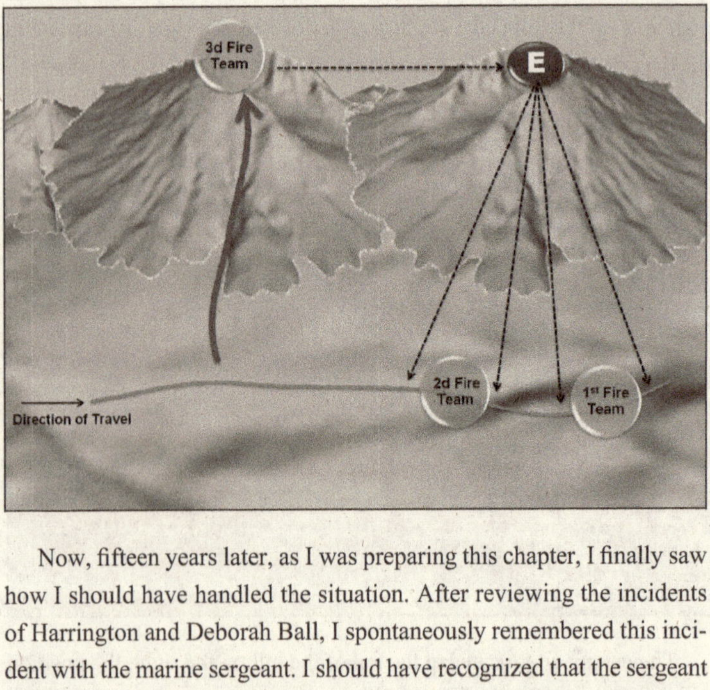

Now, fifteen years later, as I was preparing this chapter, I finally saw how I should have handled the situation. After reviewing the incidents of Harrington and Deborah Ball, I spontaneously remembered this incident with the marine sergeant. I should have recognized that the sergeant I first called on had classified his dilemma as an ambush, but, in fact, his fire team was not being ambushed. It was the 1st and 2nd fire teams that were caught in the ambush. I believe that the sergeant was applying a basic tactical doctrine: "When you are ambushed, immediately attack the main source of the fire." The rationale for the doctrine is that the enemy is usually trying to drive you into a kill zone so that if you follow the path of least resistance, you get slaughtered. Therefore, the safest tactic is to attack the enemy force directly in order to drive it off.

However, the ambush doctrine didn't apply here because the sergeant's unit, the 3rd fire team, wasn't being ambushed. The NCO was misapplying the doctrine. I should have helped him by asking for his reasoning, just as Deborah Ball helped Sean understand odd and even numbers. I could have used the sergeant's comments to diagnose the flaw in his understanding of ambushes. I could have helped him appreciate the distinction between being ambushed and being part of a unit that is being ambushed. I could have helped him think about when to apply the ambush doctrine. I blew it.

CHAPTER SIXTEEN

Helping Our Organizations

HELPING ORGANIZATIONS GAIN MORE INSIGHTS means breaking the tyranny of the down arrow in the performance equation. It means dialing back the War on Error. We'll need to restore a better balance between the arrows, between trying to reduce errors and deviations on the one hand, and increasing insights on the other (see diagram).

People working in organizations face pressure for predictability and perfection (reducing errors and deviations), which motivates managers to specify tasks and timetables as precisely as possible and to view insights as disruptive. Besides, errors are public. They're visible—easy to track and measure. Managers can show reductions in fault rates as evidence that they're doing a good job.

The simple solution is to back off, to reduce the amount of reviews and cut back on the activities designed to prevent errors. If we think of the down arrow as the brake pedal, organizations need to stop pressing so hard. It's no wonder that they're not getting very far if they are putting so much effort into slamming on their brakes.

Unfortunately, the forces of predictability and perfection will continue to exert their pressure. That's why any efforts to relax pressure on the down arrow are likely to give way over time. If I work in a large organization, I can visualize the effect on my career and my reputation if I make a blunder; I can visualize lots of ways my team and I can blunder. However, I can't visualize insights as clearly. I can't imagine what insights I might achieve if I ease up on the down arrow. Also, I can recall times when people took on the organization to advocate for their insights; most of those instances ended unhappily for the crusaders. That's why it isn't realistic to expect organizations to give up some of their cherished controls. The controls are what organizing is all about. The need for predictability and perfection motivates everyone, it structures their activities, and it is essential for managing programs, projects, and people. What can we do about it?

STRENGTHEN THE UP ARROW

We can promote countervailing forces that create pressure for insights and discoveries. This means finding ways to strengthen the upward arrow to balance the need for predictability and the fear of errors.

One idea is to set up a team of insight advocates to promote practices that encourage discoveries. The intelligence community has an office to ensure analytical integrity and critical thinking. Major corporations have departments to improve quality and reduce defects. Why not balance these with a team of insight advocates?

The team members, in addition to their regular job responsibilities, could roam around the organization collecting examples of insights and discoveries, much as in the cases I collected. The team could disseminate the best examples every month to encourage others and to show that the organization values the insights of its members. I suspect there are many good instances of insights that never get documented.

For example, years ago I was working on a U.S. Navy project to reduce the number of sailors needed on large ships. Someone on our team mentioned that I should talk to our colleague David Klinger, who had recently helped a nuclear power plant improve its handling of emergency drills. Klinger might have cut the size of the emergency response group as one of the changes he recommended, but no one was exactly sure. So I spent a few hours interviewing David and discovered that he had helped the plant cut its emergency response group from more than eighty people down to thirty-five! Part of their problem was that they were overstaffed and were interfering with each other. Klinger and I even formulated a model of how performance of a team improves as new members are added, but then suffers when it gets too large. The marginal value of the new members is smaller than their coordination costs. All this because of my accidental conversation with David. And notice that I didn't ask David if he had any insights or discoveries to share. I had a specific question—how to reduce staffing levels—that was the catalyst for our discussion.

Kurt Swogger appointed himself a one-man insight advocate at Dow Chemical when he took charge of one of its struggling divisions. Swogger started out by listening—actively listening—using a steno pad to record his conversations with the finance staff, the marketers, the scientists. "Your people actually *have* the answers. They just don't know it. And there's a piece here and a piece there that *you* have got to put together." Swogger's breakthrough came in discussions with the chemists. They had developed a new catalyst, but they didn't see where to go with it. Swogger did. His background as a chemical engineer and as a former customer enabled him to recognize how this catalyst would greatly appeal to companies using customized polymers. The team called its new product Six Days because that's how fast it delivered solutions, compared to the half year or more that companies previously had to wait. Unlike Martin Chalfie, who accidentally stumbled on the green fluorescent protein, Swogger went looking for insights.

Insight advocates could share what they learn by using stories. In September 2011, I had the good fortune of conducting a full-day workshop on insights and storytelling in Singapore with my friend Patrick Lambe, the founder of Straits Knowledge, a knowledge management company, and Shawn Callahan, the founding director of Anecdote, an Australian group that applies storytelling and narrative methods. Anecdote's tagline is "putting stories to work." Patrick used the workshop to synthesize my interest in insight and Shawn's expertise in applied storytelling.

Initially, I resisted doing the workshop because I wasn't sure what the collaboration would accomplish, but in the end I came to appreciate the link between stories and insights. The 120 cases in my research sample were all stories.

Here is one of the stories Shawn told at our session in Singapore. He had conducted a leadership workshop for a local corporation. Before the workshop he arranged for staff members to tell stories that captured good and bad company leadership. Shawn and his team wrote these stories down verbatim, just as they were spoken. They collected about one hundred stories in this way. Then Shawn asked the workshop participants to select the story that was most significant in terms of staff engagement.

The story they chose was pretty anemic, more an observation than a story. The woman who told it had said that whenever she went to her manager's office, his attention was almost totally captured by his computer screen. However, when he saw that she had entered his office, he stopped what he was doing, came over to the table in the middle of the room, sat down, and engaged her as if she were the only thing on his mind. She finished her comment by saying that she really appreciated his behavior; it made her feel valued, and no other managers in the company did it. The workshop participants discussed this narrative at length and agreed it was the most significant one in the set because it offered a simple way to improve employee engagement.

After the story program had been running for a year, the human resources director asked Shawn and his team to collect a new set of stories because the staff members were getting bored with the original set. As they collected the new stories, they found example after example of people saying things like, "Hey, I don't know what changed, but whenever I go to my manager's office these days, he/she stops what he/she is doing, comes over to the table in the middle of the office, and really focuses on me and what I have to say."

Shawn and his team were surprised by this change. They speculate that it stemmed from four causes: (1) the participants worked it out for themselves (no expert was telling them what to think), (2) they got concrete instructions on what to do in the story (stories are naturally concrete), (3) their peers all agreed it was the most important story (social proof), and (4) the story had emotion (the participants in the original workshop could *feel* the woman's reactions).

This last factor seems particularly noteworthy. The emotional content of the story resulted in action. Emotion helped to translate the insight into lasting organizational change. Shawn claims that the power of stories to produce change comes from both the emotion and the understanding.

Although I enjoy thinking about ways to set up a team of insight advocates, I confess that I am pessimistic that an organization would sustain it or any other attempt to strengthen the up arrow. A charismatic leader like Shawn Callahan can inspire organizations to change, but I'm not sure how long his narrative techniques will last once they become institutionalized. I can easily see them getting swept away in the first financial crisis and round of budget cuts, along with innovations such as the insight advocates.

Callahan disagrees, at least about the storytelling program. He told me about a skeptical leader who wanted his company, which was undergoing a financial crunch, to cancel the story-based leadership program. But then he attended a session and achieved an important insight about weaknesses in his own leadership style. At the end of the session he sent a message to his CEO saying that the program needed to be maintained and every leader must attend. He is now the CEO of the company. So Callahan believes these kinds of programs may be sustainable.

In 2012, Haydn Shaughnessy wrote a column for *Forbes* magazine describing his interviews with chief innovation officers. The Chief Innovation Office is the function needed to strengthen the up arrow and encourage insight advocates and storytelling programs. These different pieces seem to be coming together.

Nevertheless, I am still pessimistic. Too often I have seen good intentions dissolve into the flavor of the month, replaced by the next idea that excites the leadership. In contrast, the forces sustaining the downward arrow—predictability and perfection—never go out of style.

LOOSEN THE FILTERS

Instead of trying to boost the up arrow, we can tinker with the down arrow so that fewer insights get blocked. One way to counter an organization's tendency to edit things to death is to set up an alternate reporting route so that knowledge workers can publish opinion pieces that don't have to go through routine editing. This alternate publishing method would escape the filters. Unfortunately, this method may work better in theory than in practice. Workers don't want to be gadflies. They want to have an impact with the projects they do and the reports they write, and therefore they want their ideas to be the corporate view, which means going through the full review and editing process.

When I described this problem of filtering to a friend who is a three-star general, he explained the strategy he uses to jump levels. He establishes a dialogue with someone at a lower rank, a major or a lieutenant colonel, whose judgment he trusts. That way he gets access to ideas and concerns before they get filtered out. The more senior officers don't like to get cut out of the chain, but he tells them to get over it—that's how he works. If they want more direct access to him, rather than playing it safe, they will have to give him honest opinions, even if those opinions make him uncomfortable.

A former intelligence analyst, a close friend of mine, had a different idea. She agreed that her organization has too many hoops to jump through. It takes only one office, one person in the review chain, to reject a written piece. Therefore, analysts are forced to hedge and modify. They're at the mercy of the most negative links in the entire approval chain. My friend suggested a way to reduce this vulnerability: offer a court of appeals for position pieces that got rejected because they were too bold. It could be called an Oversight Group, convened as necessary to provide a path around any attempts to block an unpopular view. However, we can't go too far in loosening the filtering process. Remember, the goal is to achieve a productive balance between the two arrows, not to have either one dominate the other. An organization that entertains every possible insight that comes along will become disoriented, distracted, and dysfunctional. If middle managers in the FBI pass along every warning, they'll drown their organization in alarms. The intelligence agencies are continually receiving warnings, most of which are unfounded, but each can prompt multiple lines of speculation.

That's why an Oversight Group might work as a middle ground, not dismantling the filters but leaving them in place and providing an escape hatch for instances in which workers feel strongly about their insight. An Oversight Group could reduce the chance that a legitimate warning would be blocked by a risk-averse manager.

INCREASE ORGANIZATIONAL WILLPOWER

The organizational problem may go deeper than discouraging people from having insights or filtering out insights. In many cases, the problem isn't about having or noticing insights; it is about *acting* on them. The organization lacks the willpower to make changes. It may be blind to how urgent the situation is. Leaders know what they need to do, but they cannot muster the energy to do it. Remember how Admiral Stark had the right insight about the Battle of Taranto—that the Japanese could attack the U.S. Navy fleet at Pearl Harbor—but couldn't overcome the organizational forces that chafed against the torpedo nets he recommended as a safeguard.

Rosabeth Moss Kanter, a Harvard Business School authority on leadership and organizations, has described her frustration with this lack of corporate willpower. "I still observe executives exhibiting the same lack of courage or knowledge that undercut previous waves of innovation. They declare that they want more innovation but then ask, 'Who else is doing it?' They claim to seek new ideas but shoot down every one brought to them."

An example in the news as I write this chapter is Eastman Kodak, which has just filed for Chapter 11 bankruptcy. The company was founded in 1880. In 1888, it offered the first flexible roll film, and in 1900 it sold the $1 Brownie camera, bringing photography to the masses. Kodak continued its dominance in photography for a century. In 1976, it accounted for 85 percent of the camera sales in the United States and 90 percent of the sales of photographic film. Kodak's stock price reached a high of $95/share in 1997. However, the company went downhill from there. On January 19, 2012, the date it filed for Chapter 11 bankruptcy, its stock was selling for 36 cents per share. The company had missed the shift from conventional to digital cameras. At least that's the conventional story, a big, complacent company that lacked the insight to see how digital technology was going to spoil its business model.

The conventional story doesn't hold up very well against the facts. The digital camera was invented by none other than Kodak! Steve Sasson, an electrical engineer at Kodak, developed the first digital camera in 1975. He and his supervisor got a patent for it in 1978. Kodak didn't miss the digital revolution; it started that revolution. So this isn't a case of a company that lacked insight.

Kodak was aware of how digital technology was going to change the marketplace, but it was reluctant to give up its high profit margins from photographic film to the much smaller profits from digital cameras. The company believed it had time to make the transition. It didn't feel the urgency needed to make radical changes.

When Kodak decided to get serious about marketing digital cameras, it zoomed to number one in sales by 2005. Unfortunately, that lead was quickly eroded by aggressive competition from Nikon, Sony, and Canon. And then came cameras embedded in cell phones, making digital cameras irrelevant for consumers who wanted to capture "the Kodak moment" and send it to their family and friends a moment later.

The same frightening scenario unfolded for Encyclopedia Britannica (EB). In 1990 it had record revenues of $650 million. By 1996, its revenues were half that, and the company was sold that year for a mere $135 million. This was a sad end to a company that had produced the most prestigious encyclopedias of all time. George Washington owned a set, as did Thomas Jefferson. By the 1980s, EB could produce a complete set for only $250 but charged customers $1,500 to $2,200. The reputation of EB enabled it to attract high-profile contributors who received minimal payment but were happy to write for the prestige of having a signed article.

The core of the EB business model was the sales force, which received $500 to $600 per set. Sales reps didn't have to go door-to-door making cold calls; they made appointments with likely customers. Many of the top EB executives started as sales representatives.

Everything was looking rosy for EB, although the executives weren't sure what to make of the new information technologies, such as CD-ROM formats. In 1985, they shrugged off an offer from Microsoft to put EB on a CD-ROM, in large part because of worries about how that would affect the print version. And the sales force was adamant about preserving a business model that was working so well, especially for all the reps. Why would they want to sell the information without the books and at a cheaper price? It didn't make sense.

Did the EB executives lack insight about the potential for CD-ROMs? No, because in 1989 they offered one of the first multimedia encyclopedias on a CD-ROM. It was *Compton's Encyclopedia,* not the more respected *Encyclopedia Britannica*. It received glowing reviews, and in 1991 it was released in DOS and Mac versions. Still, the company wasn't sure how to sell it. EB offered it free to customers who purchased the print version of the *Encyclopedia Britannica* and charged others $895 for the CD-ROM. EB didn't find many who were willing to pay that much.

In 1993, Microsoft issued its own CD-ROM of Encarta, a digital version of a poorly regarded Funk and Wagnall's encyclopedia that had been sold in supermarkets as a promotional item. Encarta featured attractive graphics and video and music. Microsoft offered Encarta for free as part of its Windows suite and charged other customers only $100.

When EB sales started to slide in the early 1990s, panicky executives finally did offer a CD-ROM version of *Encyclopedia Britannica* in 1994. It was free to those who purchased a print version and $1,200 to others, but that offer was too late to catch up to Encarta. By 1996 the price of a CD-ROM *Encyclopedia Britannica* had dropped to $200. When the company was sold later that year, it was spared the ordeal of having to compete with Wikipedia, which launched in 2001.

Two companies, two giants, each dominating its market, each fated to collapse very quickly. Neither lacked insight. What they lacked was willpower. Kodak invented the first digital camera. Encyclopedia Britannica produced one of the first multimedia encyclopedias on a CD-ROM. Both companies became trapped in business models that had previously worked so well. Kodak's photographic print division resisted any shift to the lower-profit digital cameras, and EB's sales force refused to put its product on a disk. Each company needed to make changes while its original business model was still profitable, before the collapse began, but they were unable to throw away what looked like a winning hand.

With hindsight, perhaps we shouldn't judge them so severely. Neither company was aligned with the new type of business they'd have to enter. Kodak wasn't a computer hardware company and wasn't enthusiastic about entering the whirlwind of technological innovation. Kodak dominated the sales of film, but with the collapse of that market, Kodak would have had to enter the competitive world of digital cameras and would have seen that market shrink in the face of cell phone cameras. Sure, those cameras can't compete in quality, but most consumers prefer convenience over quality. I don't see a good way forward for Kodak. And I don't see one for Encyclopedia Britannica. Even if EB had managed to enter the world of compact disks, even if it could have shifted from these to the world of search terms and linkages, it would have run into Wikipedia. Neither company really belonged in the information technology business. I don't see how either company could have staved off its demise, but I still have to fault their leaders for resisting change for so long.

Kodak and Encyclopedia Britannica are sometimes cited as examples of why corporations need to continually reinvent themselves. They are compared unfavorably with companies such as the Haloid Photographic Company, manufacturer of photographic paper and equipment, which transformed itself into Xerox after World War II. They are compared with IBM, the largest information technology company in the world, which decided to sell off its personal computer, disk drive, and other hardware businesses that generated $20 billion per year in revenues, because global competition had turned these into low-profit ventures. IBM chose to become smaller but more profitable and successful, transforming itself from a maker of business machines into a global provider of information services. The company made these changes before it had to, before it entered a period of crisis.

The concept of continual transformation and reinvention sounds dynamic and exciting, and I wish I could be more enthusiastic, but I worry about the toll it takes, the confusion and coordination breakdowns from excessive reorganizations and the endless visioning retreats. It is easy to cherry-pick examples, contrasting Kodak with IBM, while ignoring companies that did try to transform themselves and got it disastrously wrong.

Mostly, I worry that the doctrine of continual transformation runs counter to the emergence of insights. Advocating for continual or even periodic transformation makes it into a routine to be performed. In contrast, insights are accidental. An organization that rigidly adheres to a doctrine of continual transformation, following the creative desperation path, is different from one that is sensitive to connections, coincidences, curiosities, and contradictions.

Organizations demonstrate willpower when they act on insights, particularly insights about their primary goals. An insight about a goal isn't about being flexible and adapting plans in order to reach the original goal. It's about changing the goal itself.

A goal insight helped to reverse the U.S. situation in Iraq during Operation Iraqi Freedom. By 2006, the conflict was going very badly for the U.S.-led coalition forces. Then U.S. Army Colonel Sean MacFarland changed direction and turned the Sunni stronghold of Anbar Province, widely viewed as a lost cause, into a pillar of strength for the coalition forces.

MacFarland started his 2006 tour in Iraq in the northwestern city of Tal Afar. For the next few months, he observed the tactics of Colonel H. R. McMaster, who instructed his young troops to treat Iraqis with dignity and respect. "Every time you treat an Iraqi disrespectfully, you are working for the enemy." This in a nutshell was the concept behind counterinsurgency. And it worked in Tal Afar. When McMaster judged he had enough local support, he attacked the insurgents in Tal Afar, drove them out, and established small outposts throughout the city. This tactic was the opposite of the existing army policy of keeping a small footprint and minimizing contact with Iraqis so as not to antagonize them.

In June 2006, the army posted the now-seasoned MacFarland to Ramadi, the largest city in Anbar Province, the largest province in Iraq, almost a third of the entire Iraqi landmass. Both the city and the province were hotbeds of insurgency. Every previous brigade assigned to Ramadi had lost a hundred soldiers or more, with nothing to show for it. This was Sunni territory, and the Sunni insurgents were furious about how the United States had overthrown Saddam Hussein and left them subordinate to the more populous Shia communities. They had scores to settle with the Shia and with the U.S. forces. A marine corps intelligence report at the time bluntly stated that Anbar Province was lost to the insurgents.

At the time he took over the army brigade in Ramadi, MacFarland found that al-Qaeda was running the city. There was no mayor, no city council, no communications, and no services. The al-Qaeda leaders made it clear that any Iraqi seen leaving the U.S. military base or city hall would be killed.

MacFarland determined to use the tactics he'd seen in Tal Afar to recover Ramadi. He didn't want to hide in a fortified position. He wanted his troops to get to know the tribal leaders. He wanted to set up outposts in Ramadi and find other ways to restrict al-Qaeda. This strategy would let American and Iraqi soldiers work side by side.

MacFarland's strategy also made the U.S. Army soldiers less predictable. Previously, they had to open the gates of their compound to start every patrol, and so al-Qaeda could watch the gates and know when the army was making the rounds. MacFarland also started a new policy of recruiting Iraqi police locally, which meant using Sunni officers so that they had a commitment to defending their homes. Recruiting increased from thirty Sunnis per month before MacFarland's policy to three hundred per month in July 2006 after the new policy started. These were bold changes that supported MacFarland's overall mission. No goal insights yet. No revelations about changing the goals he'd been given.

MacFarland also took advantage of a new political development in Ramadi: the rift between al-Qaeda and the local Sunni tribes who had been its supporters. The tribal leaders in the area resented the way al-Qaeda was taking over. In 2005, twelve tribal leaders began competing with al-Qaeda for the loyalty of the people of Anbar. Al-Qaeda then started an assassination campaign, killing half of the tribal leaders and forcing most of the others to flee the country. The United States had just stood by while this happened because the sheikhs had ties to the insurgency. The attempt of the tribal leaders to stand up to al-Qaeda failed, but it revealed the tension between the traditional leaders and the terrorist usurpers.

As MacFarland's strategy for pacifying Ramadi began working, al-Qaeda became more desperate and ramped up its terror tactics, which created more antagonism from the local Sunni tribes. For example, al-Qaeda assassinated a sheikh who refused to let its followers marry his daughters. Another provocation was an incident on August 21, 2006, when al-Qaeda killed a pro-government sheikh because he encouraged many of his tribesmen to join the police, and then hid his body in a field for three days, preventing it from receiving a proper Muslim burial.

Shortly after these outrages, tribal leaders started the Sunni Awakening, a movement to reclaim their territory from al-Qaeda. Other tribal leaders joined in. One of the leaders of the Sunni Awakening approached the U.S. Army and suggested they align their forces.

This was the critical moment for MacFarland. Officially, the offer was way out-of-bounds. His mission was to put down the Sunni insurgents and disarm them. But as he considered the changes in the political climate, and the opportunities that were opening up, he became excited. Instead of rejecting the offer, he put his faith in the Sunni Awakening and began supporting the Sunni militias that he had previously been fighting. His insight was that the U.S. coalition now had a window of opportunity in Ramadi, and he was determined to use it. From my perspective, MacFarland demonstrated a connection-type insight as he added some new and unexpected anchors to his mental model and watched the implications fly.

MacFarland's 180-degree shift outraged the Shiite leadership in Baghdad, who were caught up in a cycle of violence with Sunni insurgents that threatened to expand into a civil war. But MacFarland persisted. His goal reversal flipped Anbar Province from being a lost cause to being a haven for the United States and an inspiration for U.S. counterinsurgency tactics. Thus, on December 6, 2006, when al-Qaeda forces killed three U.S. personnel (two army soldiers and one marine) in an area controlled by a tribe aligned with the Sunni Awakening movement, the tribe responded to this insult to its honor by killing or capturing, within ten days, all the al-Qaeda insurgents involved in the attack. By early 2007, the U.S. and its Iraqi Sunni allies had regained control of Ramadi. At one point in August 2007, Ramadi went eighty consecutive days without any attacks. The tactic spread throughout the rest of Anbar Province, which became a secure zone for the U.S. military. When MacFarland reversed his original goals, he unleashed a process that had enormous repercussions.

Leaders such as MacFarland are rare. People often resist goal insights. Organizations tend to promote managers who tenaciously pursue their assigned goals. They are people who can be counted on. This trait serves them well early in their careers when they have simpler tasks to perform, ones with clear goals. That tenaciousness, however, may get in the way as the managers move upward in the organization and face more difficult and complex problems. When that happens, the original goals may turn out to be inappropriate or they may become obsolete. But managers may be incapable of abandoning their original goals. They may be seized by goal fixation.

Kishore Sengupta, a professor of information systems at INSEAD, in Fontainebleau, France, demonstrated goal fixation. In 2008, he and his colleagues published the results of their study about how managers react to misaligned goals. The Sengupta study tested hundreds of experienced managers in a realistic computer simulation. The scenario deliberately evolved in ways that rendered the original goals obsolete in order to see what the managers would do. They didn't do well. Typically, the managers stuck to their original targets even when those targets were overtaken by events. They showed goal fixation, not goal insight. They resembled Kodak and Encyclopedia Britannica, not Sean MacFarland.

If we want to help organizations that are stifling insights, perhaps we need to start by diagnosing why they are struggling. If they are so regimented and controlling and error averse that they are preventing insights, then they will need to back away from the controls and take their feet off the brakes, while also finding ways to acknowledge and encourage their insightful workers. If they are filtering out insights, then they should recalibrate their filters and also set up escape paths so that insights aren't trapped by any single point of excessive caution.

If the organization is unable or unwilling to act on insights, then its problem is hardest to remedy. Here we have moved beyond the borders of insight and into corporate mind-sets and motivations.

How will an organization know which problem or set of problems it is facing? That will take insight—self-insight at the organizational level rather than the individual level.

APPEAL TO AUTHORITY

There is one more strategy to consider—we can try to get the leaders of the organization to change the culture. If the organization is placing too much emphasis on the down arrow, on reducing errors and uncertainty, then it should suffer the consequences. It should be less innovative because it is stifling insights. As a result, the organization should be less successful than it could be. To make their organizations run better, the leaders should want to shift to a balance between the up and down arrows.

That's the theory, but why should an organization's leaders believe it? They're going to want to see the evidence. We're going to have to show them the corrosive effects of having an imbalance between the two arrows. Ideally, we would conduct a study showing what happens when an organization overloads on the down arrow. We would arrange for twenty or more Fortune 500 companies to overemphasize the down arrow and show that their performance suffers. But we know that's unrealistic. There's no way we could convince that many major corporations to enter into a study like this.

It turns out that we don't have to convince them. They convinced themselves.

In the last thirty years, American industry unintentionally performed a large-scale natural experiment. Major corporations went all in on the down arrow. The name of the experiment was Six Sigma.

Motorola developed the Six Sigma program in 1987 to counter competition from Japanese products that had consistently higher quality. Motorola and other American and European corporations were rapidly losing sales to Japanese companies. Motorola's communications sector designed a new approach—Six Sigma—to gather and use data to achieve near-perfect quality. They wanted to cut errors close to zero.

Sigma refers to variability, a measure of standard deviations from the average. Six Sigma tries to remove variability in order to avoid errors and increase predictability. Six Sigma is better than 99 percent successful. For a post office, 99 percent accuracy would mean that out of 300,000 letters, 3,000 would be misdelivered. A Six Sigma goal would be only one misdelivery. Six Sigma adherents often measure defects per million opportunities (DPMO). A one-sigma level would be 690,000 DPMO. Two sigma is 308,000 DPMO. Five sigma is 320 DPMO. Six sigma is 3.4 DPMO.

The Six Sigma program closed the quality gap with Japanese products. It took on the appearance of a cult. Highly trained Six Sigma experts were awarded black belts. Six Sigma treated the down arrow very seriously.

Many large companies such as Honeywell and 3M adopted Six Sigma. Jack Welch, the charismatic CEO of GE, publicly committed his company to Six Sigma in 1995, and the rush was on. Eventually 58 of the top Fortune 200 companies joined the Six Sigma bandwagon. They were gripped by down-arrow fever. Some early results seemed promising—sharp increases in profits.

Then the Six Sigma experiment turned sour. A 2006 article in *Fortune* reported that 91 percent of the large companies that adopted Six Sigma had failed to keep pace with the S&P 500 ever since. The reason, according to *Fortune,* was that Six Sigma got in the way of innovation. Too much energy was spent cutting defects to 3.4 per million, and not enough energy was expended developing new product ideas.

In 2007, *Business Week* described the rise and fall of Six Sigma at 3M, a company that had adopted Six Sigma in 2000 when it hired one of Jack Welch's lieutenants at GE to be its CEO. Four years later when 3M appointed a new CEO, it throttled back its commitment to Six Sigma. 3M had seen its creativity diminish under the Six Sigma regime. One senior 3M researcher complained that he had to justify his projects by listing assumptions, performing all kinds of market analyses, and documenting everything. The inventor of Post-it notes, one of 3M's best-known product development stories, concluded that his work never could have emerged if a Six Sigma process had been in place.

Some might conclude that Six Sigma should be abandoned in light of its failures, but the performance diagram asserts that we need to reduce errors and uncertainty while also increasing insights. We need to do both, not jump from one extreme to the other. Six Sigma shouldn't be abandoned; it needs to be corralled. Charles O'Reilly III and Michael Tushman have advanced the concept of an ambidextrous organization—one that pursues efficiency and reduces errors for mature products while encouraging innovation and creativity in other areas. The trick is to keep the two approaches separate. The efficiency group and the innovation group would report to the same manager but would otherwise run independently of each other, lest the culture of reducing errors and uncertainty spoil the culture of speculating and experimenting.

I have a mild concern that this ambidextrous approach diminishes the chance for insights about mature production activities and restricts their improvements to small incremental changes. But at least it is progress toward balancing the two arrows in the performance equation. Organizations need to engage in both activities at the same time but respect the different mind-sets that govern each arrow.

The appeal to authority isn't as naïve as it might appear. It builds on the hard lessons learned from the Six Sigma experiment, and it offers a path for proponents of each of the arrows. It provides a way to promote and protect insights within an organization. And it harnesses the pragmatic motivation of wanting to succeed in a competitive environment. Pragmatism, survival, and competition may be organizational forces strong enough to counter the perfection trap and the predictability trap, strong enough to resist the pull of the down arrow.

CHAPTER SEVENTEEN

Tips for Becoming an Insight Hunter

OPENING THE GATES TO INSIGHT ALSO MEANS opening ourselves to insights—being able to track and unpack them. By this point, you should be more sensitized to insights, so you are probably seeing more of them. That's the tracking part. Next comes the dissection, trying to understand how the person gained the insight. That's the unpacking part.

We can't do much unpacking of insights we spot in newspaper and magazine articles because we can't ask any questions. I ran into this problem because my research project used many valuable secondary sources. I never met Wagner Dodge, who died in 1955, relying instead on Norman Maclean's account of the Mann Gulch fire. I don't know any of the clever investors who profited from the financial collapse of the subprime housing market in 2007; Michael Lewis and *New York Times* reporter Gregory Zuckerman provided useful accounts that helped me understand the insights of several savvy investors.

I was satisfied using these secondary sources because my project was simply to dig a trench, hoping to make some discoveries. The written accounts usually had enough depth and detail for my purposes. The writers typically probed for the story of how the insights emerged. Nevertheless, I learned more when I could talk with the person who actually had the insight.

The best situation is to watch the insight unfold and then probe for more details. I remember how exciting it was sitting next to my daughter Devorah as she made her discovery about her fantasy baseball team. Those opportunities don't happen very often.

There's a middle ground between secondary sources and direct observation of insights, and that's to interview the person afterward about an insight worth examining. Many of the cases in my research sample, almost a third, came from in-depth interviews. I had a chance to interview U.S. Army General Martin Dempsey (currently the chairman of the Joint Chiefs of Staff) for several hours about how he discovered tactics for conducting urban warfare in Iraq. I was able to interview Dave Charlton about his speculation that the data on heat conduction of glass wouldn't apply to very thin glass coatings. Ginger, another friend of mine, was happy to describe her experiences struggling with the noncompete clause she had signed.

You never know when you'll spot an insight, or when you'll have a chance to interview someone, so you always have to be ready to go hunting. Here is an example.

LIZARD ZERO

In June 2010, my wife, Helen, and I went on a two-week vacation to Alaska. On the first day, waiting for the bus to pick us up at our hotel, we met another couple, Jay Cole and his wife, Carol Townsend. Pleasantries were exchanged—we were all polite because we knew we'd be spending the next two weeks together. No need to alienate people immediately. Jay mentioned that he recently had retired as curator in the Department of Herpetology and Ichthyology at the American Museum of Natural History in New York, where his research focused on lizards. I casually asked what kinds of questions he pursued, and he mentioned a curious case of a type of lizard that seemed to have regressed. Instead of sexual reproduction, it shifted back to cloning, a more primitive form of reproduction. Most of the experts rejected the possibility of regression, but it turned out to be true.

Ah, he seemed like a good candidate for my insight project. I'd be able to add a new specimen to my set of insight examples even while I was taking a vacation. Besides, how could I not be curious about these lizards? But I didn't have a chance to follow up. The vacation was too distracting; the activities were too engrossing. The first week went by, and I barely exchanged another word with Jay or Carol. Midway through the second week, I decided to be more aggressive. I approached Jay and asked if he and Carol would like to join Helen and me for dinner so that I could learn more about his research. Jay agreed.

That evening, as the people on the trip milled around the hors d'oeuvres table, I maneuvered to cut Jay out of the herd. I deflected invitations from others to join them at larger tables and steered Jay and Carol to a small table for four to make sure we didn't have any unwanted and distracting intruders. I waited for the server to come by and interrupt our conversation to take our orders.

Finally, I was in position. I went in for the kill. "Tell me about the lizards."

As Cole described it, the lizard story blended surprises, disbelief, and reverse Darwinism. In the late 1950s, a few naturalists began encountering unusual colonies of whiptail lizards in the southwestern United States and northern Mexico. The colonies consisted only of females. One of the researchers scratched his head, went to the library, checked the scientific literature, and found a 1958 publication by a Russian specialist, Ilya Darevskii, who reported finding all-female lizard colonies in Armenia. Darevskii observed 3,000 lizards of the *Lacerta* family, and none of them were male! What was going on? Were these reports accurate? Lizards rely on sexual reproduction, a male fertilizing the eggs of a female. Without any males, how did these colonies continue?

Several leading naturalists simply rejected the findings. Some doubted Darevskii just because he was Russian—the United States and the USSR were in the midst of a cold war. Even some Soviet researchers doubted Darevskii's report. The finding was too bizarre. Lizards reproduce sexually. Sexual reproduction was well established for plants, reptiles, mammals, even bacteria.

Other naturalists tried to explain the findings away. Perhaps the samples were biased. Perhaps the females had already copulated with males (who somehow wandered off or died) and stored the sperm in their oviducts for a long time. Perhaps the so-called females were hermaphroditic and reproduced through self-fertilization.

Follow-up studies cast doubt on each of these counterexplanations. One researcher dropped female lizards into crevasses to hibernate through the winter and then brought them out—they still reproduced the following spring. Histology showed that the females in these colonies were not hermaphrodites, and they didn't have any sperm stored in their oviducts.

That left an unlikely explanation: the lizards reproduced by cloning themselves. The offspring were genetic duplicates of the mothers. In all-female colonies of lizards, the embryos were developing from unfertilized eggs. The tough step is accepting that these lizard colonies were truly unisexual.

In those days before DNA analysis methods, some researchers tested the cloning hypothesis by transplanting skin tissue from one lizard to another. In sexual species, these skin transplants are rejected, but members of these unisexual (all-female) lizard species did not reject the transplants.

The cloning hypothesis didn't get firmly established until Jay Cole and others performed DNA studies. And to do that work, they needed to raise female lizards through several generations. Unfortunately, the lizards kept dying until Carol Townsend, herself a biologist and a researcher at the American Museum of Natural History (and therefore an interested party to this dinner conversation), developed methods to keep them alive.

Okay, so it was cloning. The next question—and there is always a next question (you never get to the last question because there isn't a last question)—is, How did the cloning get started? In plants, cloning sometimes happens among hybrids. In this lizard case, two related lizard species were able to mate, just as horses and donkeys can mate, and the hybrid results of such matings were sterile, just as mules are sterile. But with the lizards, at least once in Armenia and a few times in America, a specific female hybrid individual, "lizard zero," began a new line that relied exclusively on cloning to continue. This switch to cloning couldn't gradually evolve, like a giraffe's long neck or the larger prefrontal cortex of humans. It had to happen in a single generation. Otherwise the first generation of hybrids couldn't have reproduced.

So now we have the answers to the three mysteries. First, yes, there are all-female species of these lizards. Darevskii's observation was correct. Second, yes, they reproduce through cloning in the absence of sperm. Third, the cloning started with a lizard zero. Lizards from two related species mated, and their hybrid offspring, instead of being sterile, were able to give birth to clones. Cole described this sequence to me as reverse Darwinism because Darwin described the benefit of sexual reproduction for fostering lots of variation, whereas these lizards' retreat into cloning was a step in the opposite direction. Two species mated and fostered a line of unisexual offspring that showed no genetic variation. Cole tried to track down who originated the lizard zero story and narrowed it to a few of the leading investigators, but they had died, so it was too late. This bit of history was lost.

Cole had a ringside seat to this story. As a graduate student in the early 1960s interested in reptiles, he worked in the laboratory of the first American researchers to stumble upon all-female lizard colonies. In 1965, he moved to a different laboratory in Arizona and got to work on histology and genetics. In 1969, Cole accepted a position at the American Museum of Natural History in New York and began his own research program on the genetics of cloned lizards. If you go to the third floor of the American Museum of Natural History, you can see three specimens of whiptail lizards in the Hall of Reptiles and Amphibians. One is a member of an all-female species that descended from the hybrid of the other two.

Cole's story of the female lizards fits the contradiction pathway. The research community immediately fought the notion of lizards that reproduced through cloning. The researchers objected to the contradiction between their existing ideas and the field observations of these all-female lizard colonies. The naturalists who believed in cloning overcame the skeptics.

The lizard zero case was pretty straightforward. An unusual observation is resisted by the research field but then turns into an insight or a set of insights.

Hunting for insights gets more challenging when it seems highly unlikely that an individual is capable of insights or that studying a group will uncover anything new. Sometimes it just takes a different hunting strategy, as in the next case.

THE CANNY CONSUMERS

Market researchers probe for insights into their customers' decision strategies. They use surveys, questionnaires, and focus groups to learn how consumers think about products. Sometimes, though, the standard tools aren't powerful enough.

I was once on a team hired by Procter & Gamble (P&G) to help the company introduce a new detergent into the marketplace. P&G's high-quality products sold well but were a bit pricey. Procter & Gamble was missing a product at the lower end of the market. It wanted to develop one that appealed to the so-called economical homemakers—female heads of household—who refused to spend money on the high-quality detergents. Historically, P&G had neglected this part of the spectrum.

The Procter & Gamble project team had collected thousands of surveys over several years to try to understand these economical laundresses. The team had file cabinets filled with surveys. And it had a model: these consumers simply purchased the least expensive detergent in the supermarket. The P&G team was satisfied with its model. Plus, the company was getting ready to roll out the new product. The last thing anyone wanted was a delay created by outsiders like us.

Still, senior managers didn't want to miss anything. One of them had worked with my company on a previous project and had experienced the interview methods we use to get insights about tacit knowledge—subtle cues that people have trouble articulating. He directed the Procter & Gamble project team to set up a meeting with us. It was on July 6, 1994.

The P&G team explained what it needed from us, adding that we had to complete our project by the end of July. Seeing our look of concern, the P&G team members reassured us that they didn't expect us to find anything they didn't already know. After all, they had been at this for three years. We couldn't possibly learn anything new and useful in less than a month. We were being hired simply to placate the senior manager.

I walked out of the meeting feeling angry. The smugness of the P&G project leaders offended me. I was determined to make a breakthrough about the thought processes of these economical homemakers.

We had just a few weeks, and after the usual start-up delays we could schedule only two days for interviews. My three colleagues and I formed ourselves into two teams and interviewed a total of twelve homemakers for the two days. That was it. The P&G team was used to sampling hundreds of female heads of household with each survey. The P&G team couldn't believe we could get anywhere with twelve interviews. But their telephone surveys took about ten minutes. Our interviews lasted two hours. The P&G researchers had trouble believing that we could spend an entire two hours talking with one homemaker about a very simple decision.

The structure of the interview was to show each woman we interviewed a mock-up photograph of a shopping aisle in a typical supermarket and to ask her to take an imaginary walk down the aisle, stopping to inspect various brands. The photograph showed the brands but not the prices. The women would have to ask us for that information. That way, we'd know which brands they were considering. The interviewees had been prescreened; they fit the demographic of those looking for price discounts.

Our P&G sponsors watched the interviews from behind one-way mirrors. The results weren't what they expected. The homemakers were much pickier than Procter & Gamble had thought. They weren't just buying the cheapest item because if they brought home a detergent that didn't meet their quality standard, or made their family unhappy (e.g., "My clothes make my skin itch"), they'd have to stop using it, thereby wasting money. The P&G model described these homemakers as unconcerned about quality, in contrast to upscale shoppers. But we found that the economical homemakers were just as concerned about quality, plus they were demons about cost. They enjoyed finding a bargain and were proud of getting good quality at a low price. They took great satisfaction in not paying extra for fancy packaging or marketing gimmicks. They were committed to keeping the weekly shopping bill as low as possible.

Some of the women had evolved a sophisticated strategy for picking a detergent to buy. Over time they identified a set of three or four brands that their families would accept. That target set was their starting point. Then they cut coupons from the newspaper and also looked for additional bargains for any of the items in their target set when they arrived at the supermarket to do their shopping, scrutinizing the coupons and sales available. On that day, they purchased the cheapest available brand from their target set. It was important for them to have only three or four acceptable brands. If they had fewer than three possibilities, there wouldn't be enough of a chance to find a bargain that day. If they had more than four possibilities, they would have to do too much calculating.

Previously, Procter & Gamble simply had identified these consumers as having no interest in quality and as just buying the cheapest available detergent. Now P&G understood the sophisticated strategy these women used. These were not simpleminded shoppers just selecting the brand with the lowest price. They were putting in more time and analysis than any of the other customer groups.

Later, the head of the Procter & Gamble team thanked us for helping the team achieve a highly successful product rollout. We gave the team members an important insight into the decision strategy of their customers.

If we start out without much respect for the people we're investigating, we're unlikely to learn much from them. In this Procter & Gamble example, the market researchers had a dismissive attitude. Because the economical homemakers were buying inexpensive detergents, not the good stuff, the market researchers didn't take their thought processes very seriously. Our team, however, respected the homemakers. We conducted an appreciative inquiry, looking for the strengths of the homemakers.

Another example, about my grandson Koby, may help you adopt your own appreciative inquiry. Koby is a happy little boy, and he seems very serious in watching whatever is going on around him.

On his first birthday, my daughter Rebecca was reading through a questionnaire about twelve- to eighteen-month-olds. One question asked if the child could understand and respond to the question "Where's Daddy?" Rebecca had no idea but figured the answer was no. Neither she nor her husband, Matt, had ever asked Koby any questions and seriously waited for a response.

The next morning, Rebecca tried Koby out. She faced him, put her hand on her nose, and said, "Koby, this is Mama's nose." Then she put her hand on his nose and said, "This is Koby's nose." She repeated, her nose, then Koby's. Then she asked him, "Koby, where's Dada's nose?" Koby turned, crawled to his father, and grabbed his father's nose.

Rebecca and Matt were dumbfounded. They had no idea he could do these things—follow directions, know who "Dada" was, replace Matt's nose for theirs.

I am sure many babies can follow this kind of direction. This example isn't about Koby; it's about his parents, who never realized what Koby was capable of. If you don't expect much, if you don't inquire in a way that respects the intelligence of the other person, you probably won't find many insights.

The burden of appreciative inquiry, of listening or watching in a generous way rather than in a dismissive fashion, becomes much harder when you know the other person is telling you things that are wrong. How can you take that person seriously? It was only through appreciative listening that a research team I headed was able to solve a mystery that stumped British military analysts.

ONE MINUTE LEFT TO LIVE

The incident occurred during Operation Desert Storm in 1990. Michael Riley, a British naval officer, was on the cruiser HMS *Gloucester,* assigned to protect the USS battleship *Missouri* from Iraqi attacks. Riley was worried about a nearby Iraqi Silkworm missile site off the coast of Kuwait.

Toward the end of Desert Storm, Riley watched U.S. airplanes, A-6s, come streaming back from their raids, often not turning on their IFFs (Identify Friend or Foe systems), often flying directly over the *Gloucester.* Then he saw a contact—a blip on the radar screen—that froze him. In an instant, he judged that it was an Iraqi Silkworm missile fired in his direction and that he had one minute left to live.

He trained a second type of radar on the contact, confirmed that it was flying at around 1,000 feet (rather than 2,000 to 3,000 feet like an A-6), typical of a Silkworm, and ordered that it be shot down.

But how did he know? After the war, he and British military analysts studied the radar recording from this incident but could find no evidence from the initial radar returns that it really was a Silkworm missile. Silkworm missiles and A-6s are about the same size and fly at the same speed. The only differentiator is altitude, but Riley's initial radar system didn't pick up altitude. Riley reported that it was "accelerating as it came off the coast," and that's what caught his attention, but a review of the tape showed no sign of acceleration.

A research team I led had a chance to interview Riley about the incident. We spent two hours with Riley going over the critical ten seconds when he identified the threat. We replayed the recording of the same radar video he viewed and interviewed him about what he saw in it. We came up blank, the same as the engineers and analysts who studied the tape. Riley was delighted because he believed he had extrasensory perception (ESP), and his achievement suggested he was right. There didn't seem to be any other explanation.

We used a ruler to measure the first three blips on the radar screen and confirmed that once the Silkworm missile appeared on the scope, it was flying at a constant speed, not accelerating. And it would have taken at least three radar sweeps to detect acceleration, but Riley knew it was a Silkworm missile within the first five seconds, really as soon as he saw it. So he was mistaken that it was accelerating as it came off the coast.

The next morning, a member of our research team came in with the answer. The radar Riley was using picked up air contacts once they moved out to sea, after they broke free from ground clutter—the ground features that dominate the radar scope and mask aircraft. Because the Silkworm missile flew at a lower altitude than the A-6s, it didn't break free of ground clutter until it was much farther away from the coast of Kuwait than any of the A-6s Riley had seen. That's what caught Riley's attention. The first appearance of the radar blip was much farther out than he'd ever seen before. His sense of typicality had built up from watching the A-6s day after day, making it easy for him to spot an anomaly.

Once it broke out of the ground clutter, the Silkworm continued at the same speed as the A-6s. So Riley was wrong in his assessment that it was accelerating off the coast, but he was correct that what he saw *seemed like acceleration*. The missile appeared to be moving very fast when it came off the coast and then slowed down. Riley's perception was valuable even though it was wrong. If we had gone by the facts and hadn't taken Riley seriously, we would have missed it. The analysts who previously had studied the tape were too focused on the images and weren't thinking about Riley's experiences watching hundreds of A-6s on his radar scope. We brought Riley back that afternoon and explained our hypothesis. Riley had liked the idea that he was using ESP, but he reluctantly agreed that we had captured the basis for his judgment.

The Riley example shows how important it is to listen sympathetically and appreciatively to understand how a smart person can arrive at an incorrect conclusion. We don't have to believe what they are telling us. Mike Riley was wrong in the details of what he saw. Even so, we needed to listen to the clues he was providing.

There's more to the probing than appreciative listening. For example, many incidents are about insights into other people and why they behaved as they did. When trying to understand why people acted in a certain way, you might use a short checklist to guide your probing: their knowledge, beliefs and experience, motivation and competing priorities, and their constraints.

- *Knowledge*. Did the person know something, some fact, that others didn't? Or was the person missing some knowledge you would take for granted? Devorah was puzzled by the elderly gentleman's resistance until she discovered that he didn't know how many books could be stored on an e-book reader. Mitchell knew that his client wasn't attuned to narcissistic personality disorders and was therefore at a loss to explain her cousin's actions. Walter Reed's colleagues relied on the information that mosquitoes needed a two- to three-week incubation period before they could infect people with yellow fever.

- *Beliefs and experience*. Can you explain the behavior in terms of the person's beliefs or perceptual skills or the patterns the person used, or judgments of typicality? These are kinds of tacit knowledge—knowledge that hasn't been reduced to instructions or facts. Mike Riley relied on the patterns he'd seen and his sense of the typical first appearance of a radar blip, so he noticed the anomalous blip that first appeared far off the coastline. Harry Markopolos looked at the trends of Bernie Madoff's trades and knew they were highly atypical.
- *Motivation and competing priorities*. Cheryl Cain used our greed for chocolate kisses to get us to fill in our time cards. Dennis wanted the page job more than he needed to prove he was right. My Procter & Gamble sponsors weren't aware of the way the homemakers juggled the needs for saving money with their concern for keeping their clothes clean and their families happy.
- *Constraints*. Daniel Boone knew how to ambush the kidnappers because he knew where they would have to cross the river. He knew the constraints they were operating under. Ginger expected the compliance officer to release her from the noncompete clause she'd signed because his company would never release a client list to an outsider.

This checklist can suggest some ways to probe for more details that will help you unpack a story so that you can understand it better.

STICKY KEYS

I'll finish off this chapter with a very simple story about an observation from my wife, Helen, that further illustrates what I mean by unpacking an insight:

HELEN: "My front door key doesn't work well. Maybe it's the lock."
GARY: "Well, my key works fine. So it isn't the lock. Must be your key."

Gary goes to hardware store and makes a new key for Helen. Upon returning home, he tests it out. It doesn't work well. Perhaps the key duplication machine at the store was defective. Gary goes to a second hardware store and makes another key. Eventually Gary solves the problem. Helen can now get into the house without a problem.

This isn't much of a story, is it? In fact, I wouldn't call it a story at all. It's a chronology of events. A story should have a point; ideally, it should provide us with an insight. This chronology doesn't go anywhere. Lots of times, people will claim they are telling us a story, but they will merely recount a chronology of events that happened to them. Very tedious. So let's deepen this chronology by adding more details, shown in italics.

HELEN: "My front door key doesn't work well. Maybe it's the lock."
GARY: "Well, my key works fine. So it isn't the lock. Must be your key."

Gary goes to hardware store and makes a new key for Helen. Upon returning home, he tests it out. It doesn't work well. Perhaps the key duplication machine at the store was defective. Gary goes to a second hardware store and makes another key.

That key doesn't work either.

Gary tries his own key, watching carefully. It actually sticks pretty badly.

Gary squirts a lubricant into the lock. All the keys now work fine.

Eventually Gary solves the problem. Helen can now get into the house without a problem.

This version is more of a story. Not a great one, not a spellbinder, but at least it has become a story. It is an insight story. The "bad key" frame gets replaced by a "bad lock" frame. We can also give it a label: it's a garden path story. Garden path stories are how someone adopts an erroneous frame and tenaciously preserves it despite mounting evidence to the contrary. Once I got on the garden path—the erroneous belief that the problem was the key—I stayed on it. I ignored the evidence from the first hardware store and went to a second. With garden path stories, we want to know how people got on the garden path and how they get off it, if they ever do. Could I have gone to a third hardware store after my first two tries failed? I have to admit that it's possible.

In probing more deeply, we might wonder how I ever got on the garden path in the first place. It was my belief that my key was working fine, when it wasn't. Yet my key didn't suddenly start sticking. It happened gradually. So this is a "routinization of deviance" story. These are cases in which an anomalous event gets repeated often enough to become familiar so that it no longer triggers any warnings. Diane Vaughan, a professor of sociology at Columbia University, described NASA's *Challenger* tragedy as a routinization of deviance. The safety engineers initially worried when the O-rings got scorched because that wasn't supposed to happen. However, as more and more missions came back with scorched O-rings, the engineers stopped being concerned. In my case, as the lock got stickier, I came to accept the extra jiggling I needed to do as part of my practice. I got used to it, and it became the new normal for how my key worked. Over time, the jiggling increased but never enough to shake me out of my complacency.

As I unpacked this story and enjoyed the extra layers, I shared it with Helen, who had a different interpretation. She said that this story illustrated a theme of our relationship: how I don't listen to her as carefully as she would like. She had suggested a lock problem at the very beginning. Not only had I ignored her comment, I didn't even recall it until she brought it to my attention. (Note to self: work harder on appreciative listening.)

Given her reaction, should I even have told her it was a lock problem? I could have easily squirted the lubricant into the lock, given Helen any of the collection of house keys I now had, and explained that the problem was solved. So perhaps this is a story of how honorable I was to admit my own foolishness.

However, I had completely forgotten that Helen had mentioned the lock until she reminded me. This isn't a story of an honorable husband. Rather, it is the story of a stupid husband.

Whatever its other merits, this account illustrates the different layers that emerge from even very simple insights. If we can squeeze this much out of such a mundane incident, think of how much more richness we can uncover from truly challenging cases.

CHAPTER EIGHTEEN

The Magic of Insights

INSIGHTS OFTEN APPEAR LIKE MAGIC because all we see is the surprising finale, the rabbit popping out of the hat. We don't see the steps leading up to that finale, the years that the magician spent practicing, the design of the hat, the way the rabbit was smuggled on stage, the way the magician's assistant leaned forward to give us a glimpse of her cleavage at a critical instant.

Graham Wallas's model of insight helped foster the impression of insight as magic. Wallas highlighted the unpredictable flash of illumination, the rabbit suddenly drawn from the hat.

Although we may not be able to predict the exact instant when a person has an insight, the process is not as mysterious as many people think. Earlier I presented a Triple Path Model of insight with separate pathways relying on contradictions, creative desperation, and connections. Each pathway has its own means of altering the beliefs that anchor the way we understand things. This process of restructuring beliefs—changing the story we use to understand events—gives rise to the flash of illumination that Wallas described. Insights unexpectedly replace one story with a new one that is more accurate and useful.

Using the Triple Path Model, we see that Wallas wasn't wrong. He was addressing just one of the three paths—desperation—in which we reach an impasse and have to undo the flawed beliefs that are knotting our thinking. Some of our insights require us to escape from impasses. More often, we respond to contradictions and to connections.

The Triple Path Model helps resolve some of the myths surrounding insight. These myths usually fit one of the paths but not the others. Is insight a matter of breaking out of an impasse? Yes, for the creative desperation path, but not for the others. Does incubation help? The evidence is mixed, but if incubation does improve our chances for insight, it may work in different ways for the three paths. Does experience get in the way of insights? Sometimes, on the creative desperation path, but not on the others. Should we keep an open mind? Yes, on the connection path, but not always on the contradiction path, which often calls for a skeptical mind. Should we expose ourselves to lots of swirl, lots of different ideas? Perhaps for the connection/coincidence/curiosity path, but not for the others. Is insight a matter of connecting the dots and filling a gap? Not for the contradiction or the creative desperation path; on the connection path the feeling of connecting the dots is an illusion based on hindsight as we forget the non-dots and anti-dots. It's easy for us to talk past each other if we don't specify which path we're traveling.

The Triple Path Model grew out of the 120 cases that formed my project about insights. Most of the insights we've covered throughout the book clicked into place like a combination lock that pops open when we rotate the tumblers to the correct settings. Even when the answer doesn't immediately appear, we know there's a right answer someplace. Jay Cole and others used DNA evidence to confirm the lizard cloning story. Other stories from biology (the double helix model), investigations of diseases (cholera, yellow fever, ulcers, AIDS), investigations of financial fraud (Bernie Madoff), and astronomy (pulsars) have an ending. They have a solution.

Many insights, however, just feel right but don't have a correct solution or a satisfying ending. Often they are insights about people. These soft insights cannot be cleanly confirmed. Who's to say what the "right" lessons of the Battle of Taranto are? We don't know if Napoleon's tactic at Toulon was the only one that would have succeeded. The Denver Broncos might have won the Super Bowl without their plan to contain LeRoy Butler. The financial managers who profited from the collapse of the subprime housing market might not have looked so prescient if the housing bubble continued for another year—perhaps boosted by a federal rescue or by Chinese investors. Ginger's strategy for escaping the noncompete clause could have been met with a refusal to let her off the hook. The psychotherapists Mitchell Klein and Bob Barcus diagnosed and helped their clients, but there could have been other interventions that would have been as effective. Possibly the clever purchasing strategy used by some of the economical launderers was an anomaly. We can't validate these soft insights, but we should value them even if they're not as clear as the others. The Triple Path Model seems to apply to both the clear-cut and the soft insights.

Some insights are high-impact, such as Darwin's theory of evolution, Yamamoto's strategy for attacking Pearl Harbor, Einstein's theory of special relativity. Others have less impact or are even trivial. We've covered a number of everyday insights. My insight about getting my car keys to my mechanic. My insight about why my wife's house key was sticking. Cheryl Cain's insight about using chocolate kisses to get us to fill in our time cards. Devorah's insight about her fantasy baseball team. My effort to help Jimmy hit a backhand. The Triple Path Model seems to apply to both major and minor insights.

After several years of investigating the different forms of insight, I believe I have a better response than "I don't know" when people ask how they can boost the up arrow in my diagram. I think I have a richer description of insight than the Graham Wallas stages of preparation, incubation, illumination, and verification. I have some more ideas about what gets in the way of insights, why some people miss insights that are right in front of them. I have a better idea of how the down arrow in the performance equation can interfere with insights if it gets pursued too vigorously. Organizations value predictability, they dislike errors, and their single-minded emphasis on the down arrow usually ignores its effects on insights.

I also have some ideas about improving the chances of gaining insights. If we want to increase our own insights, we should know about the different paths: the contradiction path, the connection path, and the creative desperation path. Each path calls for its own methods. The contradiction path depends on our being open to surprises and willing to take them seriously even if they violate our beliefs about the ways things work. The connection path begins when we are open to experiences and ready to speculate about unfamiliar possibilities. The creative desperation path requires us to critically examine our assumptions to detect any that are tripping us up.

If we want to help others gain insights, we should listen sympathetically for flawed beliefs, like the landing signal officer trying to understand why Doug Harrington had so much trouble, or Deborah Ball wondering why Sean would think that the number "6" was both even and odd. The first part of this drill is to resist the temptation to label other people as stupid because of their "stupid" activities or assertions. Most people aren't stupid. They have somehow gotten captured by a flawed belief. The second part of the drill is to turn into a detective and try to find this flawed belief. The third part is to devise a way to help the other person discover how to replace the belief.

If we want to help organizations increase insights, we must first diagnose what is going wrong. In many cases, organizations are preventing insights by imposing too many controls and procedures in order to reduce or eliminate errors. Organizations value predictability and abhor mistakes. That's why they impose management controls that stifle insights. If organizations truly want to foster innovation and increase discoveries, their best strategy is to cut back on the practices that interfere with insights, but that will be very difficult for them to do. Organizations are afraid of the unpredictable and disruptive properties of insights and are afraid to loosen their grip on the control strategies. Never mind that these strategies work best in well-ordered, rather than complex, settings. Organizations may need to keep their desires for predictability and perfection in check.

Organizations can stifle insights by filtering them out. Sometimes proactive managers will intervene to make sure the disruptive insight is heard. But their heroic actions can get them in trouble. One suggestion is to use an Oversight Group empowered to act as a court of appeals to break through the barriers that others in the hierarchy use to block insights.

But even if filters are adjusted and disruptive insights see the light of day, organizations are often unwilling to act on them. There is little point in increasing insights or making them more visible if the organizational leaders aren't ready to make any changes.

In my search for insights I came across many popular uses of the term. A newspaper article reporting an interview with college freshmen who "offer insight" into their experience. Newsletters that are titled "insights" into whatever topic they are serving up. The Ameritel hotel chain sends out an electronic communication labeled "Innsight." These different uses of the term cater to our hunger for insights. They're making a shallow use of the term "insight," offering tidbits of information. In contrast, we've been examining the kinds of insights that provoke a shift in our understanding—an unexpected shift in the story we tell ourselves about how the world works and how to make it work better.

This shift is a discontinuous discovery; it doesn't naturally evolve from our previous beliefs. One or more of our core beliefs—the beliefs that anchor our understanding—have to be discarded or radically modified, or an additional anchor has to be added to the mix. We can trace our new story back to the original one, to see how we got to the insight, but we can't take it the other way. An insight is a leap, an unpredictable leap, to a related but different story. It catches us by surprise because it isn't the product of conscious, deliberate, deductive, or statistical inferences. Typically, the people around us don't gain the insight even if they have access to the same information we do.

The insight usually arrives as a coherent new story, not as a set of possibilities. It shifts from story (1) to the final version, story (n). Watson and Crick's discovery of the structure of DNA, and its role in genetics, came all at once, a complete package.

Other times the insight doesn't land on the final story; it transforms away from the initial story toward the resolution. The Harry Markopolos insight that Madoff was a swindler didn't immediately include the ending: that Madoff was running a Ponzi scheme. That ending came later. Markopolos's insight started him on his investigation, shifting from story (1) Madoff-as-genius, to story (2) Madoff-as-swindler, on the way to story (n) Madoff-as-master-of-a-Ponzi-scheme. The Michael Gottlieb story also didn't include the ending. The insight moved Gottlieb and the medical community from story (1) business-as-usual, to story (2) there was a new disease affecting the immune system. It didn't include story (n) the nature of AIDS and HIV.

Let's take a step back. Let's shift our gaze away from the outcome, the rabbit popping out of the hat. Instead of wondering about the rabbit itself, or even about the process of gaining insights, the method for getting the rabbit in position, let's think about how our insights change us. Because they do. Some of the famous cases changed people's careers: Chalfie, Markopolos, Gottlieb, Yamamoto, Gopnik, Darwin, and on and on.

My colleague Rob Hutton and I studied a set of successful engineers and scientists at an air force research laboratory. Once they stumbled on an important idea, they couldn't let it go. They spent months, sometimes years, pursuing it. Even when directed to move on to other projects, they would secretly continue to work on their discoveries. In many cases, they bet their careers on their insights. Their insights gripped them tightly.

Then there are insights that led to personal change: my brother Dennis, the psychotherapy patients of Mitchell Klein and of Bob Barcus, the firefighter who discovered that *he* was the attitude problem. If we keep our gaze narrowed on the insight, the rabbit, we'll just conclude that the new story, the revised set of anchors, produced a better understanding. But it does more than that. The new set of beliefs leads us to view the world differently. We have different mental equipment, different ideas about our capabilities, different priorities for what to watch and what to ignore. We have different goals. In some ways we become different people.

Now let's take another step back, away from the changes that insights stimulate. This time, let's look at the forces that drive our insights. We've examined three such forces: our tendency to notice connections, coincidences, and curiosities; our tendency to go Tilt! at inconsistencies; our tendency to weed out flawed beliefs. These are the three paths in the model, but they may be more than that. They seem to be three habits of mind.

The judgment and decision making community for the past few decades has concentrated on heuristics and biases, particularly the ways our heuristics can get us in trouble. Thus, the force of seeing connections, coincidences, and curiosities can take us too far, to the point where we see connections that aren't there. Scientists refer to these false connections as "pseudo-correlations," connections that aren't valid, and warn us about them. However, these warnings miss all that we learn by noticing connections. The warning is an example of down-arrow thinking, just worrying about the cost of a mental habit without appreciating its benefits. The heuristics-and-biases community has provided us with a stream of studies showing how our mental habits can be used against us and make us look stupid and irrational. They don't offer a balanced set of studies of how these habits enable us to make discoveries.

I see the examples in this book as a collective celebration of our capacity for gaining insights, a corrective to the gloomy picture offered by the heuristics-and-biases community. Insights help us escape the confinements of perfection, which traps us in a compulsion to avoid errors and in a fixation on the original plan or vision.

The habits of mind that lead to insights, our tendency to spot connections and coincidences, curiosities, and inconsistencies, may be important in another way. They may help us shake loose from our routine ways of thinking. With experience, we tend to arrive at more accurate beliefs and more powerful patterns and stories about how things work. If we encounter the occasional glitch, we're usually able to explain it away and hold on to our ideas. The more successful our beliefs, the harder it is to give them up. This process can result in more rigid ways of thinking.

The habits of mind we've covered—noticing new connections, going Tilt! at inconsistencies, finding flawed assumptions—may combat mental rigidity. They are forces for making discoveries that take us beyond our comfortable beliefs. They disrupt our thinking.

These forces provide the up arrow with its own energy, and they help us and our organizations break free from routines. Therefore, these forces may also be a countervailing influence against the organizational pressures for predictability and perfection, the down arrow that can block insights.

The up arrow in the performance equation has its own power. It can be slowed, but it cannot be eliminated because we are built to search for insights and to relish them. The projects that fascinate us, the hobbies that consume us, often provide us with gratifying insights. In discussions with friends and family members we frequently try to gain insights about social relationships. We can't stop ourselves from trying to complete partial patterns or probe further when we spot inconsistencies.

Having an insight is an act of creation. Each insight is the creation of a new idea that didn't exist before, often in opposition to defective ideas that formerly prevailed. No one expected this new idea to emerge, and other people who possessed the same information were unaware of its existence. No matter how much we unpack insights and demystify them, we shouldn't discard the sense that something unusual has happened, something for which we can be thankful. Something we can savor. A story we can tell to ourselves and to others.

The ancient Greeks worshipped the human capacity for insight. Scott Berkun, in examining the topic of innovation, pointed out that the Greek religious pantheon included nine goddesses who represented the creative spirit. Leading philosophers such as Socrates and Plato visited temples dedicated to these goddesses, these *muses,* who were a source of inspiration. We honor this tradition when we visit a museum, a "place of the muses," and when we enjoy music, the "art of the muses." We don't have to adopt ancient Greek religious tenets to practice the Greek reverence for insights.

Although I have tried to take some of the mystery out of insights, I don't want to diminish the awe we feel when we encounter an insight, whether it be our own or someone else's. At these times, something magical has happened. The magician walked on stage with a rabbit already hidden, but insights appear without warning, without expectation. The idea that was summoned forth did not previously exist. It wasn't even guessed at.

The magic of insights stems from the force for noticing connections, coincidences, and curiosities; the force for detecting contradictions; and the force of creativity unleashed by desperation. That magic lives inside us, stirring restlessly.

ACKNOWLEDGMENTS

I DIDN'T EXPECT TO WRITE ANOTHER BOOK, but the topic of insight grabbed hold of me and evolved into a story that I needed to tell. Fortunately, I have a wonderful set of friends and colleagues who helped me discover how to tell that story.

Several friends showed remarkable patience in reviewing a very early draft of this manuscript when I was really struggling to find a way to organize the materials: Patrick Lambe, Steve Wolf, Dennis Klein, Laura Militello, Bob Barcus, Herb Bell, Bill Ferguson, and Bud Baker. They each plowed through that material and offered very useful reactions and recommendations. In addition, Mary Beth Pringle, Diane Chiddester, and Jan Kristiansson provided valuable help with technical editing. And Michelle Welsh-Horst was not only thorough but also very tolerant of last-minute revisions during the production process.

Several people provided special assistance. Andrea Jarosz was a marvelous research assistant and colleague in conducting the research and the coding for the 120 cases of insight. Jessica Finkeldey, with her limitless patience, helped organize the material in the endnotes. Jack Whitacre generated a valuable jolt of energy and enthusiasm in running down references and information, reviewing the material, and doing everything needed to get to completion. Veronica Sanger contributed her usual miracles in taking care of production, combining rapid and accurate revisions but also catching all kinds of problems I hadn't noticed. Weekends, evenings, whenever I called on her, Veronica was there, as she has been for more than a decade.

I also thank Jay Cole, Joe Sells, Guy Vandegrift, Erik Draeger, Shawn Callahan, Marcus Mainz, Steve Fiore, Bob Swanson, Steve Swanson, Sae Schatz, Jeff Klein, Steve Deal, Malcolm Gladwell, Sandy Pierce, Kathleen Carlson, Sandie Johnson, Matt Lawlor, Ralph Keyes, and Muriel Keyes for all the advice and ideas they provided. Bill Duggan provided some of the original inspiration for studying the topic of insight. Additionally, I took advantage of Dr. Michael Gannon's research on Pearl Harbor and shallow-water torpedoes.

For the artwork and illustrations, I was fortunate to draw on Michael Fleishman to take care of the diagrams, with assistance from Anthony Leone, Jared Judson, and Veronica Sanger. Veronica Sanger prepared the diagrams of the ambush scenario. Kevin Jones generated the DNA diagram, based on a figure in James Watson's *The Double Helix*.

The Mann Gulch aerial photograph was used with permission from the U.S. Forest Service Archives. Richard Rothermel gave me permission to include his powerful diagram of the race between the smokejumpers and the fire, taken from his article "Mann Gulch Fire: A Race That Couldn't Be Won." Edward Tufte generously permitted the use of the Broad Street pump diagram from his book *Visual Display of Quantitative Information*.

I used many examples in this book; the research project was built around examples. Some of these were fleeting mentions. Of the fifty-seven major examples listed in the Story Index, four came from my previous books. The Jimmy racquetball story is from *Streetlights and Shadows*. The HMS *Gloucester* story comes from *Sources of Power: How People Make Decisions*. The Procter & Gamble story is from Crandall, Klein, and Hoffman, *Working Minds*. I've used the story of Doug Harrington and the carrier landing in *The Power of Intuition* and also in *Streetlights and Shadows*.

My agent, John Brockman, brought his legendary professionalism to bear, explaining what worked and what didn't work in a very early draft and then finding the right publisher for the book.

That publisher, PublicAffairs, assigned my book to John Mahaney, the best editor I have ever had and a true collaborator. Most of John's suggestions made immediate sense. Even when we argued about his recommendations, I trusted John's fairness and flexibility and his desire to strengthen the book.

Finally, I want to express my appreciation and love for the group that John Mahaney referred to as my brain trust: my wife, Helen, and my daughters, Devorah and Rebecca. As usual, they were the first ones to review the earliest draft and the ones I used to bounce off ideas and tough decisions. They always provided candid reactions, even when they knew I wouldn't be happy with what they had to say. They weren't worried about making me happy. They wanted to help me achieve a good outcome, which is all that I could ask from them and what I always get.

NOTES

Chapter One: Hunting for Insights

4 **"it was so smart":** Sgt. Jeffrey Flohr, personal communication, April 26, 2012.

4 **profession was out of balance:** M. E. P. Seligman, *Authentic Happiness: Using the New Positive Psychology to Realize Your Potential for Lasting Fulfillment* (New York: Free Press, 2002).

6 **the nervous system of worms:** C. Dreifus, "A Conversation with Martin Chalfie: Watching Life in Real Time," *New York Times,* September 21, 2009. See also Chalfie's Nobel Prize address, "GFP: Lighting Up Life," http://nobelprize.org/nobel_prizes/chemistry/laureates/2008/chalfie-lecture.html.

7 **he was the schnook:** A. Smith, "Interview with Martin Chalfie," 2008, http://nobelprize.org/nobel_prizes/chemistry/laureates/2008;chalfie-telephone.html.

8 **"within the living animal":** Chalfie, "GFP."

10 **"he made his numbers dance":** H. Markopolos, *No One Would Listen* (Hoboken, NJ: John Wiley & Sons, 2010), 20.

11 **"This is bogus":** Ibid., 30.

13 **"I am one sick queen":** D. Brown, "The Emergence of a Deadly Disease," *Washington Post,* June 5, 2001.

13 **"*Pneumocystis* Pneumonia—Los Angeles":** M. S. Gottlieb, H. M. Schanker, P. T. Fan, A. Saxon, J. D. Weisman, and I. Pozalski, "*Pneumocystis* Pneumonia—Los Angeles," *MMWR Weekly* 30(21) (June 5, 1981): 250–252; R. Shilts, *And the Band Played On: Politics, People, and the AIDS Epidemic* (New York: St. Martin's Press, 2007); interview transcripts (interviewees include Dr. Thomas Waldmann, Dr. James W. Curran, and Dr. Anthony S. Fauci), retrieved from the National

Institutes of Health website: http://history.nih.gov/NIHInOwnWords/docs/page_06.html.

Chapter Two: The Flash of Illumination

17 **how insight works:** G. Wallas, *The Art of Thought* (New York: Harcourt, Brace, ca. 1926).

19 **"the slow ascent of wooded hills":** Ibid., 40–41.

19 ***my undiscovered mind*:** J. Drinkwater, *Loyalties* (London: Sedgwick & Jackson, 1919).

24 **"the moment you were in before":** H. Mantel, *Wolf Hall* (New York: Picador, 2009), 189.

26 **the patterns they had learned:** G. Klein, R. Calderwood, and A. Clinton-Cirocco, "Rapid Decision Making on the Fireground: The Original Study Plus a Postscript," *Journal of Cognitive Engineering and Decision Making* 4(3) (2010): 186–209.

27 **replacing erroneous beliefs:** G. A. Klein, R. Calderwood, and D. MacGregor, "Critical Decision Method for Eliciting Knowledge," *IEEE Transactions on Systems, Man, and Cybernetics* 19(3) (1989): 462–471. Note: This article distinguished between elaborations of our situation awareness and shifts to a different interpretation of events. What I am describing here is a situation awareness shift.

27 **only three or four anchors:** G. Klein and B. Crandall, "The Role of Mental Simulation in Naturalistic Decision Making," in P. Hancock, J. Flach, J. Caird, and K. Vicente, eds., *Local Applications of the Ecological Approach to Human-Machine Systems* (Hillsdale, NJ: Erlbaum, 1995).

28 **we frame and organize events:** G. Klein, J. K. Phillips, E. Rall, and D. A. Peluso, "A Data/Frame Theory of Sensemaking," in R. R. Hoffman, ed., *Expertise Out of Context* (Mahwah, NJ: Erlbaum, 2007).

30 **interrater agreement was 98 percent:** G. Klein and A. Jarosz, "A Naturalistic Study of Insight," *Journal of Cognitive Engineering and Decision Making* 5(4) (2011): 335–351.

Chapter Three: Connections

33 **combat at sea:** G. W. Prange, D. M. Goldstein, and K. V. Dillon, *At Dawn We Slept: The Untold Story of Pearl Harbor* (New York: Penguin Group, 1991).

34 **only 40 feet deep:** Regarding the water depth of the Bay of Taranto, see "Taranto Seas, Italy," Shallow Water Hydrodynamic Finite Element Model, ISMAR, December 19, 2012, https://sites.google.com/site/shyfem/application-1/lagoons/taranto-seas.

34 **"the necessary sacrifices":** Prange et al., *At Dawn We Slept,* 11.

35 **"units based in that area":** Ibid., 40.

35 **"Naval Base at Pearl Harbor":** Ibid., 45.

36 **the Pearl Harbor commanders:** M. Gannon, "Admiral Kimmel and the Question of Shallow Water Torpedoes," talk delivered at the National Press Club, Washington, DC, November 6, 2003.

38 **the broccoli and goldfish paradigm:** B. M. Repacholi and A. Gopnik, "Early Reasoning About Desires: Evidence from 14- and 18-Month Olds," *Developmental Psychology* 33(1) (1977): 12–21.

38 **the babies generously offered:** A. Novoteny, "Awakening the Child Inside," *Monitor on Psychology* 42(1) (2011): 34–36.

40 **the variations in species he had observed:** C. Darwin, *The Origin of Species by Means of Natural Selection* (London: Penguin, [1859] 1968). See also A. Koestler, *The Act of Creation* (New York: Macmillan, 1967); and D. K. Simonton, *Origins of Genius: Darwinian Perspectives on Genius* (New York: Oxford University Press, 1999).

40 **selection to the mix:** C. Darwin, "The Descent of Man, and Selection in Relation to Sex," in *Evolutionary Writings*, edited by J. A. Secord (Oxford: Oxford University Press, [1871] 2008).

Chapter Four: Coincidences and Curiosities

45 **any obvious causal link:** This is a paraphrase of the dictionary definition of "coincidence." See *Webster's Third New International Dictionary of the English Language* (Springfield, MA: Merriam-Webster, 1986), 441.

47 **"something like this before":** D. Colligan, "The Discover Interview: Jocelyn Bell Burnell," *Discover Magazine,* November 2009, 68.

47 **"That was eureka":** Ibid.

48 **"greatest astronomical discovery of the twentieth century":** BBC Radio 4 interview, October 25, 2011.

49 **to disrupt Denver's plays:** M. Silver, "Seven Up," *Sports Illustrated,* February 1998, 50–65; P. Zimmerman, "Marked Man," *Sports Illustrated,* February 1998, 48–49.

50 **the world's first antibiotic:** R. Ogle, *Smart World: Breakthrough Creativity and the New Science of Ideas* (Boston: Harvard Business School Press, 2007). See also R. Hare, *The Birth of Penicillin* (London: Allen & Unwin, 1970).

51 **the very first Nobel Prize in Physics:** T. S. Kuhn, *The Structure of Scientific Revolutions,* 3rd ed. (Chicago: University of Chicago Press, 1962), 57–59.

51 **the first silicon solar cells:** Ogle, *Smart World.*

53 **"the illusion of knowledge":** Marshall's Nobel Prize acceptance speech, "*Helicobacter* Connections," http://www.nobelprize.org/nobel_prizes/medicine/laureates/2005/marshall-lecture.pdf.

56 **took it from there:** P. Weintraub, "The Discover Interview: Barry Marshall," *Discover Magazine,* March 2010, 66–70. See also Marshall's Nobel Prize speech.

59 **imagine that it was unhealthy:** M. Parker, *Panama Fever: The Epic Story of One of the Greatest Human Achievements of All Time—the Building of the Panama Canal* (New York: Doubleday, 2007); J. R. Pierce and J. V. Writer, *Yellow Jack: How Yellow Fever Ravaged America and Walter Reed Discovered Its Deadly Secrets* (Hoboken, NJ: John Wiley & Sons, 2005).

Chapter Five: Contradictions

62 **he published in 2002:** The Joshua Rosner and Dean Baker papers are cited in G. Morgenstern and J. Rosner, *Reckless Endan-*

germent: How Outsized Ambition, Greed, and Corruption Led to Economic Armageddon (New York: Times Books/Henry Holt, 2011).

62 **the bubble was expanding:** M. Lewis, *The Big Short: Inside the Doomsday Machine* (New York: Norton, 2010).

63 **housing prices would always increase:** Ibid., 170.

65 **some capital to get started:** Ibid., 39.

65 **to provide him with capital:** G. Zuckerman, *The Greatest Trade Ever* (New York: Broadway Books, 2009), 79.

65 **"Home prices won't even go flat":** Ibid., 96.

67 **"when your imagination is out of focus":** M. Twain, *A Connecticut Yankee in King Arthur's Court* (Berkeley: University of California Press, [1889] 1979), 436.

68 **subprime lenders were facing huge losses:** J. Birger, "The Woman Who Called Wall Street's Meltdown," money.cnn.com/2008/08/04/magazines/fortune/whitney_feature.fortune/index2.htm, 2.

68 **"going to be completely fine":** W. E. Cohan, *House of Cards: A Tale of Hubris and Wretched Excess on Wall Street* (New York: Doubleday, 2009), 39.

68 **"serious panic there":** Ibid.

69 **"who is putting it together":** Cohan, *House of Cards,* 39–40.

74 **to protect citizens from cholera:** S. Johnson, *The Ghost Map: The Story of London's Most Terrifying Epidemic—and How It Changed Science, Cities, and the Modern World* (New York: Riverhead Books, 2006); C. E. Rosenberg, *The Cholera Years: The United States in the Years 1832, 1849, and 1866* (Chicago: University of Chicago Press, 1962).

74 **how anomalies lead to breakthroughs:** T. Kuhn, *The Structure of Scientific Revolutions,* 3rd ed. (Chicago: University of Chicago Press, 1962).

76 **lengths appear too long:** W. Isaacson, *Einstein: His Life and Universe* (New York: Simon & Schuster, 2007); M. Polanyi, *Personal Knowledge: Towards a Post-Critical Philosophy* (Chicago: University of Chicago Press, 1962), 10–13.

Chapter Six: Creative Desperation

79 **when they got into trouble:** A. D. de Groot, *Thought and Choice in Chess* (The Hague, the Netherlands: Mouton, [1946] 1978).

80 **to an ocean storm:** N. Maclean, *Young Men and Fire* (Chicago: University of Chicago Press, 1992), 33.

80 **"we'll get along just fine":** Ibid., 40.

83 **"is your cross":** Ibid., 7.

84 **to join him in the ashes:** Ibid., 99.

84 **"he must have gone nuts":** Ibid., 75.

85 **pinned by a boulder:** A. Ralston, *127 Hours: Between a Rock and a Hard Place* (New York: Atria Paperback, 2004).

87 **Cheryl gained our compliance:** C. Cain, personal communication, September 6, 2010.

88 **wasn't a showstopper after all:** D. Charlton, personal communication, May 29, 2009.

90 **until his defeat at Waterloo:** W. Duggan, *Strategic Intuition: The Creative Spark in Human Achievement* (New York: Columbia University Press, 2007).

Chapter Seven: Different Ways to Look at Insight

93 **"You've seen something like this before":** D. Colligan, "The Discover Interview: Jocelyn Bell Burnell," *Discover Magazine,* November 2009, 68.

93 **development of the printing press:** J. Man, *The Gutenberg Revolution* (London: Transworld, 2002).

93 **development of mass production:** A. Hargadon, *How Breakthroughs Happen: The Surprising Truth About How Companies Innovate* (Boston: Harvard Business Press, 2003).

93 **slow hunches in innovation:** S. Johnson, *Where Good Ideas Come From: The Natural History of Innovation* (New York: Penguin Group, 2010).

97 **leading insight researchers:** R. J. Sternberg and J. E. Davidson, eds., *The Nature of Insight* (Cambridge, MA: MIT Press, 1995).

98 **System 1 thinking and System 2 thinking:** D. Kahneman, *Thinking, Fast and Slow* (New York: Farrar, Straus and Giroux, 2011).

Chapter Nine: Stupidity

118 **"not to have thought of that":** A. Koestler, *The Act of Creation* (New York: Dell, 1964), 144.

Chapter Ten: The Study of Contrasting Twins

121 **McCone went on full alert:** G. Allison and P. Zelikow, *Essence of Decision: Explaining the Cuban Missile Crisis,* 2nd ed. (New York: Addison-Wesley Longman, 1999); M. Dobbs, *One Minute to Midnight: Kennedy, Khrushchev, and Castro on the Brink of Nuclear War* (New York: Knopf, 2008); S. Kent, "A Crucial Estimate Relived," *Studies in Intelligence* 36(5) (Spring 1964): 111–119, https://www.cia.gov/library/center-for-the-study-of-intelligence/csi-publications/books-and-monographs/sherman-kent-and-the-board-of-national-estimates-collected-essays/9crucial.html.

123 **warning signs of an attack:** U. Bar-Joseph, *The Watchman Fell Asleep: The Surprise of Yom Kippur and Its Sources* (Albany: State University of New York Press, 2005).

125 **"that just ain't so":** Often attributed to Mark Twain but more likely belonging to Josh Billings. See R. Keyes, *The Quote Verifier* (New York: St. Martin's Press, 2006), 3.

125 **anomalies that contradict their beliefs:** C. Chinn and W. Brewer, "The Role of Anomalous Data in Knowledge Acquisition: A Theoretical Framework and Implications for Science Instruction," *Review of Educational Research* 63(1993): 1–49.

125 **even when these diagnoses are wrong:** P. Feltovich, R. Coulson, and R. Spiro, "Learners' (Mis)understanding of Important and Difficult Concepts: A Challenge to Smart Machines in Education," in K. Forbus and P. Feltovich, eds., *Smart Machines in Education* (Cambridge, MA: AAAI/MIT Press, 2001).

132 **through a zipperlike process:** J. D. Watson, *The Double Helix,* G. S. Stent, ed. (New York: Norton, 1980); H. F. Judson, *The Eighth Day of Creation: Makers of the Revolution in Biology* (Woodbury, NY: Cold Spring Harbor Laboratory Press, 1996); F. Crick, *What Mad Pursuit: A Personal View of Scientific Discovery* (New York: Basic Books, 1988).

132 **the double helix model of DNA:** Watson, *The Double Helix,* 123.

134 **the characteristics of genes:** Judson, *The Eighth Day of Creation,* 43–44.

135 **the implications of his own research:** Ibid., 74.

135 **she was still a few steps away:** Ibid., 148.

136 **DNA was not a helix:** Ibid., 121.

136 **Wilkins showed to Watson:** Ibid., 113–114.

136 **"could arise only from a helical structure":** Ibid., 98.

136 **anyone else at the time:** Ibid., 100–102, 142–143.

136 **the way Watson did:** Crick, *What Mad Pursuit,* 68.

136 **how DNA replicated:** Judson, *The Eighth Day of Creation,* 115.

137 **not the helix shape for DNA:** Crick, *What Mad Pursuit,* 62.

137 **he stumbled upon it himself:** Judson, *The Eighth Day of Creation,* 149.

138 **he stumbled upon the complementary pairings:** Ibid., 59.

138 **to be a false trail:** Ibid., 111–112, 118–119.

138 **to recover more easily than others:** Ibid., 146.

Chapter Eleven: Dumb by Design

140 **looking for an opportunity:** The term "Indian" is used here because, according to C. C. Mann, *1491: New Revelations of the Americas Before Columbus* (New York: Knopf, 2005), that is how most indigenous peoples describe themselves.

140 **RESCUING JEMIMA:** The account of the kidnapping is from M. Faragher, *Daniel Boone: The Life and Legend of an American Pioneer* (New York: Holt Paperbacks, 1992), 131ff.

147 **his links to conservative friends:** E. Pariser, *The Filter Bubble: What the Internet Is Hiding from You* (New York: Penguin, 2012).

Chapter Twelve: How Organizations Obstruct Insights

153 **they claim to want creative ideas:** J. S. Mueller, S. Melwani, and J. A. Goncalo, "The Bias Against Creativity: Why People Desire but Reject Creative Ideas," *Psychological Science* 23(2012): 13–17.

158 **setting up bureaucratic procedures:** S. Lyall and N. Kulish, "Crisis at BBC Brought Rules, Then a Failure," *New York Times,* November 14, 2012.

163 **filtering out the Phoenix memo:** *The 9/11 Commission Report: Final Report of the National Commission on Terrorist Attacks upon the United States* (New York: Norton, 2010); M. H. Bazerman and M. D. Watkins, *Predictable Surprises: The Disasters You Should Have Seen Coming, and How to Prevent Them* (Boston: Harvard Business School Press, 2004).

163 **described the incident:** J. Davis, "Why Do Bad Things Happen to Good Analysts?," in R. Z. George and J. B. Bruce, eds., *Analyzing Intelligence: Origins, Obstacles, and Innovations* (Washington, DC: Georgetown University Press, 2008).

163 **"battering away at the Wall":** Ibid., 162.

164 **to prepare for Midway:** C. L. Symonds, *The Battle of Midway* (New York: Oxford University Press, 2011).

165 **how individuals and teams detected anomalies:** D. Snowden, G. Klein, Chew L. P., and C. A. Teh, "A Sensemaking Experiment: Techniques to Achieve Cognitive Precision," in *Proceedings of the 12th International Command and Control Research and Technology Symposium, Naval War College, Newport, Rhode Island, June 19–21, 2007,* CD-ROM (Washington, DC: U.S. Department of Defense Command and Control Research Program, 2007).

166 **to discount anomalies and disruptive insights:** G. Klein, "Critical Thoughts About Critical Thinking," *Theoretical Issues in Ergonomics Science* 12(3) (2011): 210–224.

169 **"subversive of its basic commitments":** T. Kuhn, *The Structure of Scientific Revolutions,* 3rd ed. (Chicago: University of Chicago Press, 1962), 5.

Chapter Thirteen: How Not *to Hunt for Insights*

172 **evaluation pressure interferes with insight:** T. M. Amabile, *The Social Psychology of Creativity* (New York: Springer-Verlag, 1983).

172 **banana that was beyond his reach:** W. Koehler, *The Mentality of Apes* (New York: Harcourt, Brace, 1925). Koehler's original 1917 publication (revised in 1921) was in German. This first English report was translated from the 1921 edition.

173 **each turn is supposed to pivot on a dot:** T. C. Kershaw and S. Ohlsson, "Multiple Causes of Difficulty in Insight: The Case of the Nine-Dot Problem," *Journal of Experimental Psychology: Learning, Memory, and Cognition* 30(1) (2004): 3–13.

174 **How is that possible?:** R. J. Sternberg and J. E. Davidson, "The Mind of the Puzzler," *Psychology Today* 16 (1982): 37–44.

175 **reviews a lot of this research:** A. Dietrich and R. Kanso, "A Review of EEG, ERP, and Neuroimaging Studies of Creativity and Insight," *Psychological Bulletin* 136(5) (2010): 822–848.

175 **water in one of the jars:** A. S. Luchins and E. H. Luchins, *Rigidity of Behavior: A Variational Approach to the Effect of Einstellung* (Eugene: University of Oregon Books, 1959), 108ff.

178 **"no more research is needed on this topic":** S. Ohlsson, "Information-Processing Explanations of Insight and Related Phenomena," in M. T. Keane and K. J. Gilhooly, eds., *Advances in the Psychology of Thinking,* vol. 1 (London: Harvester Wheatsheaf, 1992), 37.

178 **"to explain this type of insight":** Ibid., 39.

Chapter Fourteen: Helping Ourselves

183 **time to leave them behind:** D. Klein, personal communication, September 2010.

185 **ways to increase creative turbulence:** S. Johnson, *Where Good Ideas Come From* (New York: Riverhead Books, 2010).

186 **sprinkled in some additional bathrooms:** W. Isaacson, *Steve Jobs* (New York: Simon & Schuster, 2011), 431.

187 **"in not making useless combinations":** A. Damasio, *Descartes' Error: Emotion, Reason, and the Human Brain* (New York: Penguin, 2005), 188.

188 **to believe specific claims:** R. J. Sternberg, H. L. Roediger III, and D. F. Halpern, *Critical Thinking in Psychology* (New York: Cambridge University Press, 2007).

188 **to explore new ideas:** G. Klein, "Critical Thoughts About Critical Thinking," *Theoretical Issues in Ergonomics Science* 12(3) (2011): 210–224.

188 **THE PAUSE THAT REFRESHES:** R. M. Olton and D. M. Johnson, "Mechanisms of Incubation in Creative Problem Solving," *American Journal of Psychology* 89(4) (1976): 617–630.

189 **chances of having an insight:** R. A. Dodds, T. B. Ward, and S. M. Smith, "A Review of Experimental Literature on Incubation in Problem Solving and Creativity," in M. A. Runco, ed., *Creativity Research Handbook,* vol. 3 (Cresskill, NJ: Hampton Press, 2012). See also C. A. Kaplan, "Hatching a Theory of Insight: Does Putting a Problem Aside Really Help? If So, Why?" (PhD diss., Carnegie-Mellon University, 1989).

190 **listed some possibilities:** S. Hélie and R. Sun, "Incubation, Insight, and Creative Problem Solving: A Unified Theory and a Connectionist Model," *Psychological Review* 117(3) (2010): 994–1024.

Chapter Fifteen: Helping Others

195 **running a business with her cousin:** M. N. Klein, personal communication, January 3, 2012.

195 **some material to read:** E. D. Payson, *The Wizard of Oz and Other Narcissists: Coping with the One-Way Relationship in Work, Love, and Family* (Royal Oak, MI: Julian Day Publications, 2002).

196 **their strained relationship:** R. Barcus, personal communication, November 15, 2012.

199 **make their own discoveries:** G. Klein, *Streetlights and Shadows: Searching for the Keys to Adaptive Decision Making* (Cambridge, MA: MIT Press, 2009), 169–170.

201 **he nailed all his landings:** G. Klein, *The Power of Intuition* (New York: Currency/Doubleday, 2003), 224–226; Klein, *Streetlights and Shadows,* 173–175.

202 **"not what you think":** E. Green, "Building a Better Teacher," *New York Times*, March 7, 2010, http://www.nytimes.com/2010/03/07/magazine/07Teachers-t.html?pagewanted=all&_r=0.

Chapter Sixteen: Helping Our Organizations

209 **smaller than their coordination costs:** D. W. Klinger and G. Klein, "Emergency Response Organizations: An Accident Waiting to Happen," *Ergonomics in Design* 7(3) (1999): 20–25.

209 **"*you* have got to put together":** J. Liedtka, R. Rosen, and R. Wiltbank, *The Catalyst: How YOU Can Become an Extraordinary Growth Leader* (New York: Crown Business, 2009), 178.

211 **chief innovation officers:** H. Shaughnessy, "4 Reasons Why 2013 Will Be the Year of the Innovator," *Forbes,* December 31, 2012, http://www.forbes.com/sites/haydnshaughnessy/2012/12/31/4-reasons-why-2013-will-be-the-year-of-innovation/.

214 **"shoot down every one brought to them":** R. M. Kanter, "Innovation: The Classic Traps," *Harvard Business Review* 84(11) (2006): 72–83.

214 **a company that lacked insight:** "The Last Kodak Moment?," *The Economist,* January 14, 2012, 63–64; "Kodak to Stop Making Cameras," *Dayton Daily News* (Associated Press), February 10, 2012, A7.

215 **unfolded for Encyclopedia Britannica:** J. Bosman, "After 244 Years, Encyclopedia Britannica Stops the Presses," *New York Times,* March 13, 2012.

218 **"working for the enemy":** T. E. Ricks, *The Gamble: General David Petraeus and the American Military Adventure in Iraq, 2006–2008* (New York: Penguin Books, 2009), 60.

220 **he was determined to use it:** Ibid.; J. F. Sattler, "Fallujah—The Epicenter of the Insurgency," in T. S. McWilliams and K. P. Wheeler, eds., *Al-Anbar Awakening,* vol. 1: *American Perspectives: U.S. Marines and Counterinsurgency in Iraq, 2004–2009* (Quantico, VA: Marine Corps University Press, 2009).

221 **react to misaligned goals:** K. Sengupta, T. K. Abdel-Hamid, and L. N. Van Wassenhove, "The Experience Trap," *Harvard Business Review* 86(2) (February 2008): 94–101.

222 **Six sigma is 3.4 DPMO:** P. S. Pande, R. P. Neuman, and R. R. Cavangh, *The Six Sigma Way: How GE, Motorola, and Other Top Companies Are Honing Their Performance* (New York: McGraw-Hill, 2000).

223 **developing new product ideas:** B. Morris, "New Rule: Look Out, Not In," *Fortune,* July 11, 2006, http://money.cnn.com/2006/07/10/magazines/fortune/rule4.fordtune/index.htm.

223 **a Six Sigma process had been in place:** "At 3M, a Struggle Between Efficiency and Creativity," *Business Week,* June 10, 2007, http://wwwbusinessweek.com/stories/2007-06-10/at-3m-a-struggle-between-efficiency-and-creativity.

223 **innovation and creativity in other areas:** C. A. O'Reilly III and M. L. Tushman, "The Ambidextrous Organization," *Harvard Business Review* 82(4) (April 2004): 74–81.

Chapter Seventeen: Tips for Becoming an Insight Hunter

229 **to give birth to clones:** C. J. Cole, personal communication, June 16, 2010, October 14, 2010, October 22, 2010, and November 9, 2010.

232 **successful product rollout:** B. Crandall, G. Klein, and R. R. Hoffman, *Working Minds: A Practitioner's Guide to Cognitive Task Analysis* (Cambridge, MA: MIT Press, 2006), 216–218.

235 **even though it was wrong:** G. Klein, *Sources of Power: How People Make Decisions* (Cambridge, MA: MIT Press, 1998), 35ff.

236 **reduced to instructions or facts:** A. Gawande, "Big Med," *New Yorker,* August 13–20, 2012, 55.

238 **a "routinization of deviance":** D. Vaughan, *The Challenger Launch Decision: Risky Technology, Culture, and Deviance at NASA* (Chicago: University of Chicago Press, 1996).

Chapter Eighteen: The Magic of Insights

246 **air force research laboratory:** G. Klein and R. Hutton, *The Innovators: High-Impact Researchers at the Armstrong Laboratory Human Engineering Division* (AL/CF-FR-1995–0027 for Wright-Patterson AFB, OH: Armstrong Laboratory, 1995).

248 **who represented the creative spirit:** S. Berkun, *The Myths of Innovation* (Sebastopol, CA: O'Reilly Media, 2010), 5.

248 **"art of the muses":** Ibid.

STORY INDEX

Chapter Fourteen: Helping Ourselves

Chapter Fifteen: Helping Others

Chapter Sixteen: Helping Our Organizations

Chapter Seventeen: Tips for Becoming an Insight Hunter

INDEX

Erik Owen

Gary Klein, PhD, a senior scientist at MacroCognition LLC, was instrumental in founding the field of naturalistic decision making. Dr. Klein received his PhD in experimental psychology from the University of Pittsburgh in 1969. He spent the first phase of his career in academia and the second phase working for the government as a research psychologist for the U.S. Air Force. The third phase, in private industry, started in 1978 when he founded Klein Associates, a research and development company that had grown to thirty-seven employees by the time he sold it in 2005. He is the author of *Sources of Power: How People Make Decisions*; *The Power of Intuition*; *Working Minds: A Practitioner's Guide to Cognitive Task Analysis* (with Beth Crandall and Robert Hoffman); and *Streetlights and Shadows: Searching for the Keys to Adaptive Decision Making*. Dr. Klein lives in Yellow Springs, Ohio.

Praise for

Seeing What Others Don't

'No one has taught me more about the complexities and mysteries of human decision-making than Gary Klein.'

Malcolm Gladwell

'Gary Klein is a living example of how useful applied psychology can be when it is done well.'

Daniel Kahneman, author of *Thinking, Fast and Slow*

'Gary Klein is one of the most influential psychologists today.'

Wired

'Gary Klein has done it again in this brilliant and 'insightful' book about how our minds work, and how we make sense of what is in front of us. *Seeing What Others Don't* is remarkable on many fronts, most notably in being a pleasure to read, easy to grasp, and filled with practical advice. It is a must-read for parents, executives, coaches, teachers, healthcare providers, military tacticians, security and intelligence agents, and those in dangerous and stressful positions. Read, learn, perform and enjoy.'

Eduardo Salas, PhD, President of the Human Factors and Ergonomics Society

'A brilliant discourse on a fascinating subject. It's written in a crisp, fluent, Gladwellish way and the pages flit by.'

John McLaren, chairman of the Barchester Group in *Management Today*

'An insightful guide to insight. Klein is particularly good at identifying the barriers to getting there, such as allowing ourselves to be gripped by flawed beliefs, a lack of experience, being too passive, or following excessively rigid modes of reasoning. *Seeing What Others Don't* moves our understanding of insight forward by significantly adapting the story we have about it.'

Julian Baggini in *Financial Times*

'Vital to marketers looking to understand consumers' needs, *Seeing What Others Don't* explains insight, how it works, why insights still matter, what triggers them and shows tips on how to nurture them.'

Ed White, Editor of *Contagious*

'Klein takes us on a fascinating journey from medical breakthroughs to military strategy. He analyses why IT systems are 'dumb by design', and examines how Darwin started to understand evolution as well as how Crick and Watson discovered DNA. *Seeing What Others Don't* rattles along with pace and flair while being appropriately enough, packed with insight.'

Engineering & Technology UK

'A timely contribution asking professionals, who want to improve their personal and professional performance, to think radically about the way they work.'

The Actuary

'Klein writes with such passion and conviction that you quickly come round to his point of view. The book explains how companies and organisations try to block these insights, no matter how much they may claim the contrary. Using over 100 real examples to illustrate his theories, he explains the concepts clearly and concisely. Klein also reveals how you can boost those insights – that's invaluable advice.'

Rating 9/10 Book of the Week - *Social Bookshelves*